Fabian Gerson

High Resolution E. S. R. Spectroscopy

Chemical Topics for Students

1 Edited by
Wilhelm Foerst
and Helmut Grünewald

Fabian Gerson

High Resolution E.S.R. Spectroscopy

Translated by Express Translation Service, London

John Wiley & Sons Ltd 1970
Verlag Chemie

TITLE OF THE ORIGINAL GERMAN PUBLICATION:
HOCHAUFLÖSENDE ESR-SPEKTROSKOPIE
EDITED BY WILHELM FOERST AND HELMUT GRÜNEWALD

With 43 figures and 14 tables

LIBRARY OF CONGRESS CATALOG CARD NUMBER: 68 - 9357
ISBN 0 471 29701 1 John Wiley & Sons Ltd

Printed in Germany

Composition: Mitterweger KG, Heidelberg. Printer: Colordruck, Heidelberg
Creation of Jacket: Hanswalter Herrbold, Opladen
Illustrations: Gert Nemela, Schwetzingen

Editors' Preface

Modern chemistry can no longer be taught and studied simply as organic. inorganic, and physical chemistry. First, the boundaries between these areas have become fluid; and, secondly, neighbouring fields – in particular, some in medicine and biology – have been added but cannot be arranged in this classification.

It is difficult to allow for this variety of matter in a normal chemistry course, for it is hardly possible to devote special lectures at a University or Technical College to each of the important partial areas. On the other hand, it would be excessive to demand that even an advanced student should read extensive monographs in order to obtain his first introduction to a subject.

We have therefore created the series of "Chemical Topics" whose volumes are intended to be, as it were, primers that impart the most important facts from a single field. Those who wish to pursue a theme further can do so from the literature cited.

The "Chemical Topics" are intended to be read in conjunction with the basic textbooks. They extend the coverage of the latter, increase the penetration, and lead the reader to the latest position in a given field. It is hoped to treat all the important partial areas during the coming years, creating in the course of time a kind of encyclopaedia of modern chemistry.

W. Foerst
H. Grunewald

Preface

The term "high-resolution spectroscopy" may be used for that branch of electron spin resonance (ESR) which deals with organic radicals in solution, since such radicals generally give rise to very narrow hyperfine lines. Organic radicals with aromatic character possess a relatively high stability and have recently been subjected to detailed ESR studies of great importance, particularly for theoretical organic chemistry.

ESR spectroscopy of aromatic radicals offers a unique way of determining the spin population at various sites in the molecule, from which certain deductions can be made as to the probability of the residence of unpaired electrons on these sites, and, in a more far-reaching manner, as to the geometry and the electronic structure of the molecule. For an interpretation of the experimental data, theoretical models of radical ions were needed that in some respects extend beyond the previous qualitative pictures. A strong and very fruitful correlation between experiment and theory has emerged therefrom. Thus, the theoretical models explain the appearance and the finer details of the spectra, while the spectra can verify the predictions of the models. This is particularly important with aromatic compounds for which these models, which are based on quantum mechanics but are necessarily simplified, can be used in a general manner.

The present book is intended as an introduction for organic chemists into the electron spin resonance spectroscopy of aromatic radicals. It will be confined to radical-ions, for which a great deal of data is now available, because these can be easily prepared from the corresponding diamagnetic parent compounds by standard methods. However, the theoretical treatment is basically also valid for neutral or "odd" radicals. Since interest is focused on the spin population, the discussion is generally restricted to the hyperfine structure of the ESR spectra. The line-widths will be considered only insofar as this is necessary in connection with the resolution, the analysis of the hyperfine structure, and the explanation of certain anomalies in the spectra. The theoretical treatment will not be extended to the line-width and the g-factors. Numerous books and reviews dealing with the ESR spectroscopy of organic radicals in particular[1,2] and with ESR methods in general[3-15,301,302,304,323] can be recommended for further study.

Except for the Appendix to Part I, (A.1.1 and A.1.2), the description of the MO methods presupposes but a nodding acquaintance with this approach, though a knowledge of the Hückel model is assumed. The Appendix mentioned requires a much deeper knowledge, but one can omit it without making Part II difficult. In connection with the examples mentioned in Part II, many results published after 1966 could not be included. It should also be added that the coupling constants listed represent only a limited selection of the published experimental data[16,303].

Remarks on the English Edition

The English edition is a translation of the original German version: „Hochauflösende ESR-Spektroskopie, dargestellt anhand aromatischer Radikal-Ionen", Verlag Chemie 1967. Unfortunately, for reasons, which were outside the author's control, this book could not be published before summer 1970.

The original text has been slightly altered and completed in several sections. A new appendix on „Anisotropy Effects" has been added. By mistake, decimal places in the numbers listed in tables and diagrams remain as commas (instead of decimal points). However, this mistake should not give rise to any confusion.

F. Gerson

Table of Contents

1. General Aspects

1.1. Fundamentals of electron spin resonance spectroscopy

Resonance condition. The electron has a non-classical, intrinsic angular momentum called spin. According to the principles of quantum mechanics, only one of the components of the spin in a given direction (z) can be measured accurately apart from its magnitude. The spin is characterized by the quantum number $S = \frac{1}{2}$, its component along the z-axis being characterized by the quantum number $M_S = \pm\frac{1}{2}$. The electron thus has two spin states differing in M_S, which are described by convention as follows:

$M_S = +1/2$: Spin up (↑) or α
$M_S = -1/2$: Spin down (↓) or β.

The spin of the electron gives rise to a magnetic moment μ_E, whose z-component μ_E^z can assume only two values corresponding to the spin quantum numbers $M_S = \pm\frac{1}{2}$. M_S and μ_E^z are related by the equation:

$$\mu_E^z = -M_S \cdot g_E \cdot \beta_E = \begin{cases} -(+1/2)\, g_E\, \beta_E & \text{for } M_S = +1/2 \\ -(-1/2)\, g_E\, \beta_E & \text{for } M_S = -1/2, \end{cases} \qquad (1)$$

where β_E is the Bohr magneton, a constant having the value of 9.2733×10^{-21} erg/gauss, used here as a micromagnetic unit; g_E is a dimensionless number whose value for free electrons is 2.0023. In this case then, $\mu_E^z \approx \pm\beta_E$. On the other hand, the g_E factor (or value) for electrons in atoms or molecules may be different. Owing to the negative charge of the electron, μ_E^z and M_S differ in sign [17].

Since the z-axis can be chosen at will, it is convenient to identify it with some specific direction such as that of the applied magnetic field. The component μ_E^z of the magnetic moment is then responsible for the behaviour of the electron in the magnetic field.

In the absence of an external magnetic field, the two spin states with $M_S = +\frac{1}{2}$ and $M_S = -\frac{1}{2}$ are degenerate, i.e., they have the same energy.

When a magnetic field H is applied along the z-axis, it interacts with the magnetic moment μ_E of the electron, and the spin states are no longer degenerate (Zeeman effect). The energy of interaction E is given by the formula

$$E = -\mu_E^z \cdot H = +(M_S\, g_E\, \beta_E) \cdot H. \qquad (2)$$

The energy level E_1 ($= -\frac{1}{2} \cdot g_E \beta_E H$) of the spin state with $M_S = -\frac{1}{2}$ is lower than the energy level E_2 ($= +\frac{1}{2} \cdot g_E \beta_E H$) of the spin state with $M_S = +\frac{1}{2}$. The energy difference,

$$E_2 - E_1 = g_E \beta_E H, \tag{3}$$

that is, the gap arising between the originally degenerate spin states when the magnetic field is switched on, is proportional to the intensity H of this field.

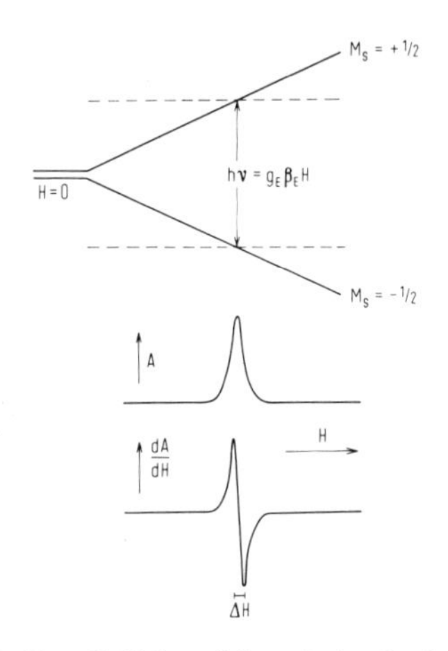

Figure 1: Top: Splitting of the spin levels of an electron in an applied magnetic field of intensity H.
middle: ESR signal obtained at constant frequency ν and variable field strength H;
bottom: derivative of the absorption intensity A with respect to H, as a function of H.

Transitions from one Zeeman level to the other (i.e. between E_1 and E_2), in which the electron changes its spin state (the spin is said to flip) occur when the system is exposed to an electromagnetic radiation with a resonance frequency ν. This frequency is determined by the resonance condition

$$h\nu = g_E \beta_E H, \tag{4}$$

where h is Planck's constant (6.624×10^{-27} erg.sec); ν thus depends on H, and the proportionality constant between them, γ_E, amounts to 2.802 MHz. gauss^{-1}, when $g_E = 2.0023$.

$$\gamma_E = \frac{\nu}{H} = \frac{g_E \beta_E}{h} \tag{5}$$

To satisfy the resonance condition, one can vary ν and/or H. For technical reasons, the frequency ν is kept constant and the field strength H is varied to bring it to the value at which the resonance condition is fulfilled. One generally uses microwaves with a frequency of 9500 MHz, which requires a field of 3400 gauss when the g_E factor is about 2. The corresponding energy is of the order of 1 cal/mole, and thus is much smaller than that encountered in UV and IR spectroscopy.

Figure 1 (bottom) shows that the signal is recorded as a derivative of the intensity A of the absorption with respect to the field strength H, and as a function of H, i.e. it is dA/dH vs. H. This way of recording the spectrum depends on the modulation used to amplify the absorption signal (cf. Section 1.3). The line-width ΔH (in gauss) or $\Delta\nu$ (= $\gamma_E \Delta H$ in MHz) is generally taken as the abscissa distance between the maximum and the minimum of the dA/dH curve.

Relaxation. When the resonance condition is fulfilled, transitions can occur either from the lower to the upper energy level or the other way round. Transitions $E_2 \longrightarrow E_1$ and $E_1 \longrightarrow E_2$ signify energy emission and absorption, respectively. Whether the system emits or absorbs, depends on the direction in which the greater number of transitions take place. Since the transition probability is the same for both $E_1 \longrightarrow E_2$ and $E_2 \longrightarrow E_1$, the decisive factor must be the population of the two Zeeman levels. Absorption can thus be observed only when the population n_1 of the lower level is greater than the population n_2 of the upper level. According to Boltzmann's distribution law

$$\frac{n_2}{n_1} = \exp\left[-(E_2 - E_1)/kT\right] = \exp\left(-g_E \beta_E H/kT\right) \qquad (6)$$

(where k is the Boltzmann constant, 1.3805×10^{-16} erg.deg^{-1}), there must be a slight population excess (n_1-n_2) at the lower level in a magnetic field H [for H = 3400 gauss and T = 300 °K, $(n_1\text{-}n_2)/(n_1 + n_2) \approx 10^{-3}$]. Since the populations are equal in the absence of a field, the application of a field can bring about the distribution, specified by eq. (6), only if a number $(n_1\text{-}n_2)/2$ of electron spins (called hot spins) flip from level E_2 to level E_1. This presupposes an energy exchange with the environment which is not possible when the electron is fully isolated. This energy exchange, known as spin-lattice relaxation, also prevents the disappearence of the field-induced population excess (n_1-n_2) during the irradiation of the system. When one irradiates (with a resonance frequency ν) a system of spins which is distributed in a field H over the levels E_1 and E_2 according to eq. (6), one finds that transitions $E_1 \longrightarrow E_2$ are more numerous than transitions $E_2 \longrightarrow E_1$, because $n_1 > n_2$. In the absence of the spin-lattice relaxation ensuring the return of the hot spins from E_2 to E_1, the populations n_1 and n_2 would soon become equalized. The time in which the number of the hot spins is decreased by a factor of 1/e is called the relaxation time T_1. It is a measure of the willingness of the lattice to receive excess spin energy, thereby re-establishing the Boltzmann distribution.

Since the relaxation time determines the life-time Δt of a spin state M_S, it is related to the uncertainty ΔE of the Zeeman levels E_1 and E_2 by Heisenberg's relation

$$\Delta E \cdot \Delta t \approx h/2\pi, \tag{7}$$

and therefore, it affects the line-width of the ESR signals.

Besides spin-lattice relaxation, there is another mechanism determining the line-width, namely spin-spin relaxation. This embraces all processes of interaction between spins, which will be discussed in Section 1.3 in more detail (cf. also Appendix A.1.3 and A.2.3). The spin-spin relaxation time is denoted by T_2, and the line-width can then be expressed as follows*):

$$\Delta \nu \propto \frac{1}{T_1} + \frac{1}{T_2}. \tag{8}$$

The line-width is thus effectively determined by the more efficient relaxation mechanism i.e. by that one which has the shorter characteristic time T_1 or T_2. Exceptionally short relaxation times can lead to such extensive line broadening that no ESR signal is detected at all. On considering the role of the spin-lattice relaxation in establishing a Boltzmann-type equilibrium (eq. (6)), one finds that moderately active spin-lattice relaxation mechanisms (moderate T_1 values) and as inefficient spin-spin relaxation mechanisms as possible (very high T_2 values) are needed to obtain narrow lines.

To sum up, ESR absorption spectra are brought about by an external magnetic field of intensity H, an electromagnetic radiation of frequency ν, and spin-lattice and spin-spin relaxation mechanisms characterized by times T_1 and T_2 respectively. As for the system under study, paramagnetic properties are required.

Paramagnetism of organic radicals. Systems with more than one electron can be characterized by two spin quantum numbers, S^{tot} and M_S^{tot}, resulting from the quantum numbers S and M_S of the constituent electrons[17]. For a given value of S^{tot} there are $(2\,S^{tot} + 1)$ possible values of M_S^{tot}. The spin state of the system is described as singlet, doublet, triplet, etc. state, according to whether $(2\,S^{tot} + 1) = 1,2,3$, etc. This number is known as the multiplicity of the state. Singlet states are diamagnetic, those of higher multiplicity being paramagnetic.

---

*) For organic radicals in solutions $T_1 \gg T_2$, so that eq. (8) is usually applied as $\Delta\nu \propto 1/T_2$.

In the ground state of stable organic compounds, not only the inner orbitals, but also the bonding orbitals are occupied, each by two electrons differing in spin quantum number M_S and known as paired electrons. The pairing refers to their spins and is frequently denoted as ↑ ↓. The state of systems in which all electrons are paired off in this way is singlet ($S^{tot} = 0$; $M_S^{tot} = 0$). Consequently, most organic compounds are in the diamagnetic singlet ground state*) and lie outside the domain of ESR spectroscopy.

If, however, a change in the number of electrons yields an unpaired one in the molecule (cf. Section 1.2), a paramagnetic doublet state results ($S^{tot} = \frac{1}{2}$; $M_S^{tot} = \pm\frac{1}{2}$), which can give rise to an ESR signal. The unpaired electron determines the magnetic properties of the doublet state, and so the ESR signal is a consequence of this unpaired electron. Organic molecules in the doublet state are known as radicals. Their paramagnetism is caused almost exclusively by the spin of the unpaired electron, so that their g_E values differ only slightly from that of free electrons (2.0023). The differences, which are due to spin-orbit coupling, exceed 5×10^{-3} only when the radical contains elements from higher Periods; in the case of hydrocarbon radicals[78] they are less than 5×10^{-4}. ESR spectroscopy is the method of choice for detecting organic radicals. Its sensitivity exceeds by several powers of ten that, for instance, of susceptibility measurements with a magnetic balance, and it permits the detection of extremely small amounts of radicals in diamagnetic substances. However, the detection of free radicals would not by itself justify such an expensive and complicated apparatus, if one could not also make important deductions as to the electronic structure of the radicals by observing the hyperfine splitting of the ESR signals.

Hyperfine structure. A well resolved ESR signal of a radical in solution may consist of more than a hundred lines, so that it is proper to speak of an ESR spectrum. This complexity of the signal, known as its hyperfine structure is determined by the interaction between the unpaired electron and the magnetic nuclei in the radical, i.e. nuclei with a non-zero spin quantum number I, analogous to the electron spin quantum number S of an electron.

A description of the behaviour of a nucleus in a magnetic field H involves the component of the nuclear magnetic moment μ_N in the direction (z) of the field (μ_N^z). By virtue of

---

*) In the excited state, organic compounds can also exhibit higher multiplicities, e.g. by the promotion of an electron from a doubly occupied orbital to a vacant orbital of higher energy, giving rise to two singly occupied orbitals. In accordance with Pauli's principle, not only a singlet, but also a triplet state is possible for such a singly excited system.

$$\mu_N^z = +M_I g_N \beta_N , \tag{9}$$

μ_N^z depends on the spin quantum number M_I. This relationship is fully analogous to that between μ_E^z and M_S in the case of the electron (cf. eq. 1). M_I can assume (2I + 1) values, namely -I, (-I + 1), (-I + 2), ... + I.

The g_N values are dimensionless and characteristic of the type of the nucleus. For the most important isotopes they are positive and vary between 0.1 and 6. Some such values are listed in Table 1, together with the corresponding quantum numbers I and M_I.

Table 1: Characteristic constants of some nuclei[18].

Isotope	Natural abundance %	Spin quantum numbers I	M_I	g_N-Factor
^1H	99,98	1/2	± 1/2	5,5854
^2H=D	0,016	1	0, ± 1	0,8574
^6Li	7,43	1	0, ± 1	0,8219
^7Li	92,5	3/2	±1/2, ±3/2	2,1707
^{12}C	98,89	0	0	0
^{13}C	1,108	1/2	±1/2	1,4043
^{14}N	99,63	1	0, ±1	0,4036
^{15}N	0,365	1/2	± 1/2	-0,5661
^{16}O	99,96	0	0	0
^{17}O	0,037	5/2	± 1/2, ± 3/2, ± 5/2	-0,7572
^{19}F	100	1/2	±1/2	5,2546
^{23}Na	100	3/2	± 1/2, ± 3/2	1,4774
^{31}P	100	1/2	±1/2	2,2610
^{32}S	99,26	0	0	0

Isotope	Natural abundance %	Spin quantum numbers		g_N-Factor
		I	M_I	
^{33}S	0,74	3/2	±1/2, ±3/2	0,4285
^{35}Cl	75,4	3/2	±1/2, ±3/2	0,5473
^{37}Cl	24,6	3/2	±1/2, ±3/2	0,4555
^{39}K	93,08	3/2	±1/2, ±3/2	0,2606
^{41}K	6,91	3/2	±1/2, ±3/2	0,1430

The above values show that, amongst the four most common nuclei in organic compounds, ^1H and ^{14}N are magnetic, whereas ^{12}C and ^{16}O are not.
The nuclear magneton β_N (= 5.0493 x 10^{-24} erg/gauss) is smaller than the Bohr magneton β_E by a factor of 1836, which is the ratio between the mass of the proton and that of the electron. The values of μ_N^z are therefore also correspondingly smaller than μ_E^z, and, provided that $g_N > 0$, they have the same sign as M_I, owing to the positive charge of the nucleus.
In a strong magnetic field H the interaction between the unpaired electron and a magnetic nucleus comes into play as a small perturbation δE to the Zeeman levels E_1 and E_2 of the electron spin, this perturbation being made up of two terms:

$$\delta E = (\delta E)_{aniso} + (\delta E)_{iso} \,. \tag{10}$$

The anisotropic term $(\delta E)_{aniso}$ represents the classical dipole-dipole interaction, which depends on the relative positions of the magnetic moments of the unpaired electron and the nucleus, μ_E and μ_N. In the case of single crystals, this interaction provides valuable information as to the geometry of the radical [310], but in amorphous and polycrystalline substances, it causes such line broadening that the hyperfine structure can rarely be resolved. On the other hand, in the case of liquids, where the molecular motion continuously changes the relative positions of the magnetic moments, the dipole-dipole interactions average out except for a small residue that depends on the viscosity of the medium (cf. Section 1.3 and Appendix A.1.3). It contributes to the line-width $\Delta\nu$ but not to δE, since the time average $(\delta E)_{aniso}$ is now zero. The hyperfine structure of radicals in solution is therefore due exclusively to the isotropic or direction-independent Fermi contact term [19] $(\delta E)_{iso}$. In a strong magnetic field in the z-direction, this term can be expressed as

$$(\delta E)_{iso} = -\frac{8\pi}{3} (\mu_E^z \mu_N^z) \rho'(0) \,. \tag{11}$$

The great importance of the Fermi term in theoretical chemistry is due to the fact that it varies not only with $\mu_E^z \mu_N^z$ but also with the electron spin density $\rho'(0)$ at the nucleus*) (cf. Section 1.5 where the concept of spin density is discussed). Using eqs. (1) and (9) for μ_E^z and μ_N^z, one obtains

$$(\delta E)_{iso} = +\frac{8\pi}{3} g_E \beta_E g_N \beta_N (M_S M_I) \rho'(0). \tag{12}$$

Eq. (12) shows that, for positve g_N and $\rho'(0)$, the levels E_1 and E_2 of the unpaired electron are stabilized when M_S and M_I have the opposite sign, being destabilized when they have the same sign.

The diagrams in Figure 2 refer to radicals in which the unpaired electron interacts only with one nucleus having a spin quantum number I of $\frac{1}{2}$ or 1 (e.g. ^1H or ^{14}N, respectively). The levels E_1 and E_2 of the electron spin states with $M_S = -\frac{1}{2}$ and $M_S = +\frac{1}{2}$ each split into two or three sub-levels corresponding to the two quantum numbers $M_I = -\frac{1}{2}$ and $+\frac{1}{2}$ of the proton or to the three quantum numbers $M_I = -1$, 0, and + 1 of the ^{14}N nucleus. For $M_S = -\frac{1}{2}$, sub-levels with $M_I > 0$ lie below and those with $M_I < 0$ lie above the unperturbed level E_1. For $M_S = +\frac{1}{2}$ (level E_2), the reverse is true. The sub-levels with $M_I = 0$ always have the same position as the unperturbed levels E_1 and E_2, since the perturbation $(\delta E)_{iso}$ vanishes in this case [cf. eq. (12)].

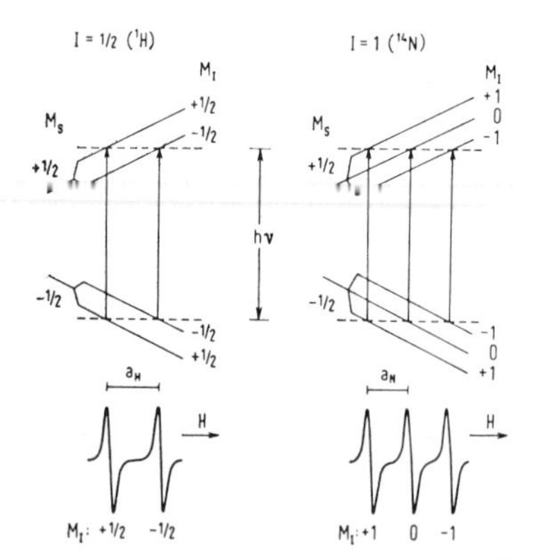

Figure 2: Splitting of an ESR signal by the hyperfine interaction between the unpaired electron and a nucleus having a spin quantum number $I = \frac{1}{2}$ (left) or I = 1 (right). The ESR hyperfine components are shown below as derivatives of the absorption intensity with respect to the field strength (cf. Figure 1): a_H and a_N denote the coupling constants for the interaction of the unpaired electron with a proton and with a ^{14}N nucleus, respectively.

---

*) The nucleus is chosen as the origin of the coordinate system.

The selection rules

$$\Delta M_S = \pm 1; \; \Delta M_I = 0 \tag{13}$$

state that only those transitions are allowed which occur between spin states with the same quantum number M_I. Consequently, two hyperfine lines are detected in the spectrum resulting from the interaction of the unpaired electron with a proton ($I = \frac{1}{2}$; $2I + 1 = 2$), and three hyperfine lines when the electron interacts with ^{14}N ($I = 1$; $2I + 1 = 3$), as can be seen in Figure 2. The separation between the adjacent lines gives the coupling constants of the nuclei in question (a_H and a_N in gauss or oersted). These are independent of the field and characteristic of the nucleus-electron interaction in the radical.

Like the corresponding interactions $(\delta E)_{iso}$, the coupling constant of a given nucleus depends only on the spin density $\rho'(0)$ at the nucleus. In the case of the proton, the constant a_H can be calculated from the equation

$$a_H = K_H \cdot \rho'(0), \tag{14}$$

where

$$K_H = \frac{4}{g_E \beta_E} \left| \frac{(\delta E)_{iso}}{\rho'(0)} \right| = 4 \cdot \frac{8\pi}{3} g_N \beta_N |M_S M_I| \tag{15}$$

$$= 2{,}3626 \times 10^{-22} \text{ erg/gauss (or gauss·cm}^3\text{)}.$$

The hyperfine lines resulting from the interaction between the unpaired electron and one magnetic nucleus have the same intensity, because, owing to the small value of $(\delta E)_{iso}$, the differences between the populations of the sub-levels belonging to the same state M_S and different states M_I are extremely small [except at very low temperatures; cf. eq. (6)]. Diagrams similar to those in Figure 2 can also be drawn up for radicals containing more than one magnetic nucleus, by splitting the levels E_1 and E_2 in succession according to the interactions with the nuclei in question. When some of the nuclei are equivalent, one derives typical hyperfine structures, such as those shown in Fig. 3 for the case of two nuclei with spin quantum number $I = \frac{1}{2}$ (protons) and $I = 1$ (^{14}N nuclei). These structures originate from the degeneracy of certain spin configurations of the equivalent nuclei; e.g. in the case ot two equivalent protons A and B, they arise from the degeneracy of the following two configurations (cf. left-hand diagram in Figure 3):

$$M_I(A) = +\tfrac{1}{2}, \; M_I(B) = -\tfrac{1}{2}$$

and

$$M_I(A) = -\tfrac{1}{2}, \; M_I(B) = +\tfrac{1}{2}$$

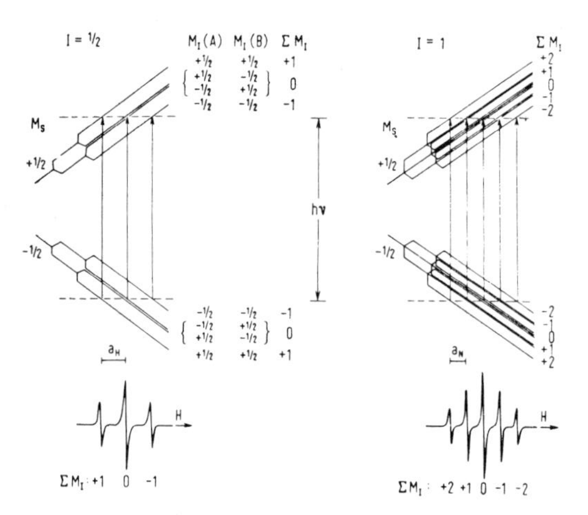

Figure 3: Splitting of an ESR signal by the hyperfine interaction between the unpaired electron and two equivalent nuclei A and B having spin quantum numbers I = $\frac{1}{2}$ (left) or I = 1 (right).

When the spin configurations of the equivalent nuclei are p-fold degenerate, the transitions between the attending levels have the same energy and result in a hyperfine line with a relative intensity p. The total number of lines is greatly reduced in this manner. For two equivalent nuclei with I = $\frac{1}{2}$, the spectrum consists of three lines with an intensity distribution of 1:2:1, and for two equivalent nuclei with I = 1 it consists of five lines with an intensity ratio of 1:2:3:2:1. In general, n equivalent nuclei with spin quantum number I give rise to (2nI + 1) equidistant hyperfine lines. In the case of protons and other nuclei with I = $\frac{1}{2}$ (e.g. ^{13}C, ^{19}F, and ^{31}P), the ESR spectrum contains n + 1 lines. For nuclei with I = 1 the corresponding number is 2n + 1 (e.g. ^{14}N and ^{2}H = D; cf. Section 1.4). The following scheme shows the number of lines and their intensity distributions for n = 1 to 6 and nuclei with I = $\frac{1}{2}$, as well as for n = 1 to 4 and nuclei with I = 1. The scheme for I = $\frac{1}{2}$ is represented by Pascal's triangle, since the intensity distribution of the n + 1 lines belonging to nuclei with I = $\frac{1}{2}$ is binomial.

I = 1/2	n	I = 1:
1	n = 0	1
1 1	1	1 1 1
1 2 1	2	1 2 3 2 1
1 3 3 1	3	1 3 6 7 6 3 1
1 4 6 4 1	4	1 4 10 16 19 16 10 4 1
1 5 10 10 5 1	5	
1 6 15 20 15 6 1	6	

If a radical contains 1,2,3, . . . k sets, each consisting of $n_1, n_2, n_3 \ldots n_k$ equivalent nuclei with spin quantum numbers $I_1, I_2, I_3 \ldots I_k$ then the total number N of the lines is given by

$$N = (2n_1I_1 + 1)(2n_2I_2 + 1)(2n_3I_3 + 1)\ldots\ldots.(2n_kI_k + 1). \quad (16)$$

This can be illustrated on the example of the radical-ion of 1,3,6,8-tetraazapyrene:

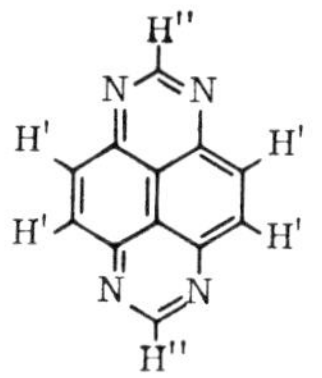

nucleus	^{14}N	$^{1}H'$	$^{1}H''$
n_k	4	4	2
I_k	1	1/2	1/2
$(2n_kI_k + 1)$	9	5	3

$N = 9 \times 5 \times 3 = 135$

1.2. Preparation of aromatic radical-ions

Radicals were described in Section 1.1 as doublet state paramagnetic molecules that contain an unpaired electron. Organic radicals tend to change into singlet state diamagnetic molecules in which all the electrons are paired and which represent the normal bonding state of organic compounds. This tendency is the reason for the high reactivity and short lifetime of organic radicals. In fact, the latter can generally be stabilized only by restricting their motion e.g. by "freezing" them into a solid matrix of a diamagnetic material. However, a result of this freezing is that the dipole-dipole interactions between the unpaired electron and the magnetic nuclei do not average out, and the ESR hyperfine lines are extensively broadened (cf. Section 1.1). For the best possible resolution of the isotropic hyperfine structure, the spectra must therefore be recorded in solution. This presupposes that the radical in question is sufficiently stable under these conditions. Sufficient stability has indeed been found for aromatic radicals, in which the unpaired electron occupies a delocalized π-orbital.

Aromatic π-radicals are divided into neutral radicals and radical-ions, differing in some theoretical aspects, chemical and physical properties, and methods of preparation. In both types one π-electron is not paired, so they contain an odd number of π-electrons. However, the situation is different as regards the number of the sp^2-hybridized atoms (π-electron centres) over which the unpaired electron can delocalize. Except for some heterocyclic compounds like pyrrole, each centre contributes one electron to the π-system in aromatic molecules. Neutral radicals thus generally have the same number of centres as π-electrons. The number of centres, too, is therefore odd, and these radicals are known as "odd radicals". By contrast, radical-ions contain an even number of centres. In anions the number of π-electrons is one more, and in cations it is one less, than the number of centres.

Odd radicals are of historical importance, since for decades they were the only stable radicals known. The classic example is triphenylmethyl, described by Gomberg[20] in 1900. Aromatic radical-ions were discovered only later[21-23], and their detailed investigation was made possible only by ESR spectroscopy[24,25]. Radical-ions form when neutral aromatic compounds acquire or give up a π-electron. The acquisition of an electron (reduction) leads to a radical-anion, while the release of an electron (oxidation) yields a radical-cation. The uptake of an electron is generally accompanied by liberation of energy, known as the electron affinity. For aromatic hydrocarbons, the amount of energy varies between 0.1 and 1 eV (2-23 kcal/mole)[26]. To release an electron, on the other hand, energy must be supplied to the molecule. The amount needed is given by the ionization potential, which is between 7 and 10 eV (160-230 kcal/mole)[26-28] for aromatic hydrocarbons. It follows, therefore, that the uptake of an electron should be much easier than its release from the same compound, and indeed many more radical-anions are known than radical-cations. In accordance with the molecular orbital (MO) models of alternant hydrocarbons (cf. Section 1.6), the same benzenoid hydrocarbons usually have both high electron affinities and low ionization potentials[26]. It is thus often possible to prepare with ease radical-cations from such compounds, which also form particularly stable radical-anions (e.g. anthracene, tetracene, and perylene). The preparation of aromatic radical-ions can be effected by numerous chemical, electrolytic, and photolytic methods.

Chemical methods. Reduction with alkali metals is the oldest and most frequently used chemical method for the preparation of radical-anions[22,24]. It is carried out in a solvent such as 1,2-dimethoxyethane (DME) or tetrahydrofuran (THF) in the absence of air and moisture.

$$\mathrm{Ar} + \mathrm{Me} \longrightarrow \mathrm{Ar}^{\ominus} + \mathrm{Me}^{\oplus}$$

The formation of radical-anions is accompanied by the appearance of a bright colour.

This method can be used with almost all aromatic hydrocarbons[29,30] and most heterocyclic compounds[31-33]. Care is needed to stop the reaction in time, because anions $\mathrm{Ar}^{\ominus}$ may otherwise pick up another π-electron to form diamagnetic dianions

$Ar^{\ominus\ominus}$ [34,35,79].

$$Ar^{\ominus} + Me \longrightarrow Ar^{\ominus\ominus} + Me^{\oplus}$$

Fortunately, the equilibrium

$$Ar + Ar^{\ominus\ominus} \rightleftharpoons 2\,Ar^{\ominus}$$

lies well to the right, in general, and the dianion $Ar^{\ominus\ominus}$ is present in a notable concentration only after all the molecules Ar have been reduced to $Ar^{\ominus}$.

Temperatures of below −100 °C and a permanent contact with the alkali metal are needed with compounds that are difficult to reduce, giving unstable radical anions and hardly any dianions. The reducing agent in such cases is a K-Na alloy in a mixture of DME and THF [36]. On the other hand, easily reducible compounds such as quinones, ketones, and nitro derivatives can be converted into the radical-anions by mild reagents such as glucose, sodium dithionite, and zinc [25,37-39]. Reaction between aromatic hydrocarbons and alkali or alkaline-earth metals in liquid ammonia also leads to radical-anions [40].

No chemical method of such general applicability is available for the preparation of radical-cations. Some hydrocarbons form radical-cations on dissolution in concentrated sulphuric acid [41-44]:

$$2\,Ar + 3\,H_2SO_4 \longrightarrow 2\,Ar^{\oplus} + 2\,HSO_4^{\ominus} + H_2O + H_2SO_3$$

In such solutions deep-coloured radical-cations are present in small amounts besides diamagnetic proton adducts. This method has been used successfully with anthracene, tetracene, perylene [45], pentacene [45], coronene [46], biphenylene [47,48], acepleiadylene [29,49], acenapth[1,2-a]acenaphthylene [49], 9-methyl- and 9,10-dimethylanthracene [50,51], pyracene [52], 1,2,3,6,7,8-hexahydropyrene [53], 15,16-dimethyldihydropyrene [54], and acepleiadiene [29,55]. Radical-cations are formed in concentrated sulphuric acid also from some heterocyclic compounds such as 1,3,6,8-tetraazapyrene [58] and systems of the 1,4-dithiin type [56,57]. A number of hydrocarbons are oxidized to radical-ions by antimony pentachloride in methylene chloride [59]:

$$2\,Ar + SbCl_5 \longrightarrow 2\,Ar^{\oplus} + 2\,Cl^{\ominus} + SbCl_3.$$

This method sometimes works even when conc. sulphuric acid fails, e.g. with pyrene, dibenzo[a,c]triphenylene, $\Delta^{9(9')}$-bifluorene, and tetraphenylethylene. Although naphthalene gives a radical-cation in this manner, it can be shown to be a dimer by the hyperfine structure of its ESR spectrum [59].

Another chemical method for the preparation of radical-cations is a reaction in which diprotonated quinones or aromatic diazo-compounds are reduced with zinc or sodium dithionite[60].

$$\mathrm{Ar + 2\,H^{\oplus} \longrightarrow ArH_2^{\oplus\oplus}}$$

$$\mathrm{2\,ArH_2^{\oplus\oplus} + Zn \longrightarrow 2\,ArH_2^{\dot{\oplus}} + Zn^{\oplus\oplus}}$$

Electrolytic methods. Though polarographic methods had long been known for the reduction and oxidation of aromatic compounds[61-63], only after 1960 were electrolytic methods first used to prepare radical-ions for ESR studies[64-66,200-239]. Radical-ions have been prepared in relatively polar solvents such as acetonitrile (ACN), N,N-dimethylformamide (DMF), or dimethylsulphoxide (DMSO) with tetraalkylammonium perchlorate as electrolyte. The radical-anions are formed on the surface of a mercury pool used as a cathode for the reduction. In the case of oxidation, the radical-cations arise at a platinum gauze anode. Generally, calomel electrodes are used in both cases as counter-electrodes. The first radical-ions prepared by these methods were the anion of nitrobenzene[64] and the cation of p-phenylenediamine[65].

After these early results, the electrolytic methods were widely applied with great success notably by Fraenkel and coworkers[67-72]. The electrolytic method has advantages over chemical reduction with alkali metals in the following four respects:

1. The radical-anions are produced continuously and directly in the ESR measuring cell, and so a sufficiently high constant concentration of the short-lived species is frequently ensured.
2. The reduction can be done at the lowest possible potential, so that further reactions that require a higher potential, such as the uptake of a second electron, are avoided.
3. It is possible to use relatively polar solvents, in which radical-anions do not associate with their gegenions, and this makes for a simpler ESR spectrum (cf. Appendix A.2.2).
4. One can prepare radical-anions of aromatic compounds with substituents such as halogens, which cannot be converted into the radical-anions by alkali metals.

However, electrolytic reduction fails under standard conditions (at room temperature with ACN, DMF or DMSO as a solvent) if the compound to be reduced has a low electron affinity (i.e., when it is difficult to reduce), and if the stability of the resulting radical-ion is such as to require low temperatures. Until recently, one could turn only to chemical reduction in such cases, but a new possibility is now offered[73,74] by the electrolytic production of radical-anions in liquid ammonia at −80 °C. This method has already given radical-anions even of compounds (e.g. 1,3-butadiene and pyridine) that polymerize on contact with alkali metals[73,167].

Relatively few accounts have been published of radical-cations prepared by electrolytic oxidation[65,75-77,200,285]. The compounds giving rise to such species in this manner are readily oxidizable amino-substituted hydrocarbons.

Photolytic methods. While chemical and electrolytic methods have a fairly general applicability, irradiation with UV light has been used to prepare aromatic radical-ions in solution only in isolated cases. Thus, radical-anions have been obtained from some ketones[37] and nitro compounds[38] in ethanol containing a large amount of sodium ethoxide. Also, high-intensity UV irradiation of 1,2,4,5-tetramethylbenzene and hexamethylbenzene in sulphuric acid solution gives rise to the corresponding radical-cations[80].

1.3. Optimal conditions for recording ESR spectra

When the ESR spectrum points to a radical-ion in solution, the first task is to extract information from it as to the electronic structure of the radical-ion. The more numerous the readily indentifiable hyperfine lines, the more satifactory will be this information in amount and certainty. The number of observable hyperfine lines depends on the resolution (i.e. the line-width) and the intensity of the spectrum (i.e. signal-to-noise ratio). There are two types of factors determining the quality of ESR spectra: instrumental factors, which depend on the spectrometer and its setting, and internal factors, which depend on the nature of the sample.

Instrumental factors. The ESR spectrometers made available in the past few years are relatively easy to operate and can be used by chemists lacking experience in electronics. It is therefore not necessary to discuss here the construction of these spectrometers, and the interested reader is referred to some other books[5,7,10,330] or any instruction manual. The following description will be confined to the basic components (see Figure 4).

Figure 4 shows that this apparatus differs from UV and IR spectrometers mainly by incorporation of a magnet in addition to usual components (source of radiant energy, absorption cell, and detector), since an external magnetic field is a prerequisite for the absorption of energy in the present case, as in NMR spectroscopy. The field can generally reach 10000 gauss and is freely adjustable. Between the pole

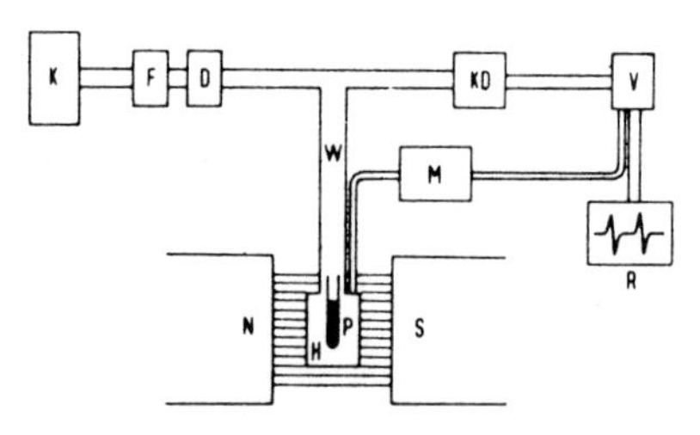

Figure 4: Basic structure of an ESR spectrometer.
N and S = poles of an electromagnet; H = resonant cavity; P = sample, W = wave-guide; K = klystron; F = ferrite insulator; D = attenuator; KD = crystal detector; V = amplifier; R = recorder; M = modulator.

pieces (10-30 cm across), it is homogeneous to 1 part in 10^5-10^6. In the homogeneous field region a resonant cavity is placed, which accommodates the sample and is connected to all other components. The quality of this resonator is measured by its ability to store energy supplied to it. Energy losses arise particularly when the sample includes a polar liquid. These losses can be diminished by reducing the cross-section of the sample holder, or by changing the cell shape to fit the electrical and magnetic lines of force in the resonant cavity. The second means is preferred, because the signal strength depends on the cross-section of the absorption cell.

Energy is supplied to the cavity through a wave-guide, the source being a valve called a klystron, capable of emitting electromagnetic radiation in a narrow range of the microwave region. Two types of klystrons emitting in the Q-band (ca. 8mm) and in the X-band (3 cm) are used in ESR spectroscopy. According to the resonance condition [eq. (4)], the Q-band klystron (ν = 36000 MHz) requires stronger fields (H) than the X-band klystron (ν = 9500 MHz) so that a more favourable Boltzmann distribution [eq. (6)] is obtained with it, together with a stronger absorption. In studies of organic radicals in solution, however, this advantage is cancelled out by the necessity of using narrower sample holders in conjunction with Q-band klystrons, because with samples of the same dimensions, the energy loss of the resonant cavity is greater at a wavelength of 8 mm than at 3 cm.

Interposed between the klystron and the wave-guide are an attenuator, to regulate the power input, and a ferrite insulator, to protect the klystron from reflected radiation. This radiation reaches the detector, a crystal diode, via a T-shaped bridge called a 'magic Tee'. The bridge can be so adjusted that no radiation reaches the detector if no absorption of microwaves occurs in the resonator. To raise the sensitivity of the crystal diode, however, one must supply a small amount of energy to it, even in the absence of absorption. This is done by throwing the bridge slightly out of equilibrium, i.e. introducing what is known as leakage. The noise which then appears, even in the absence of an ESR absorption, is the usual background of the signal: it is partly the intrinsic noise of the detector and partly the frequency noise of the klystron. The signal-to-noise ratio is improved by modulating the magnetic field, generally with a 100 kHz component. The modulated signal appears as the first derivative of the intensity with respect to the field strength and is registered as such by a pen recorder. Strong signals can also be observed on an oscilloscope screen.

One can easily establish optimal working conditions of recent spectrometer models by trying out some settings of the microwave intensity, the modulation amplitude, the amplification etc. It is sufficient to draw attention here to undesirable effects exerted by saturation and overmodulation on the resolution and/or the intensity of the hyperfine lines. Since the appearance of ESR signals is due to the absorption of microwaves, it is tempting to use high-intensity radiation in the hope of obtanining stronger signals. However, such irradiation weakens and distorts the hyperfine lines, the phenomenon involved being known as saturation. Saturation occurs, because strong irradiation intensifies the transitions between the electron spin energy levels E_1 and E_2 to such an extent that the small excess population $(n_1 - n_2)$ of the lower level is reduced faster than it can be re-established by spin-lattice relaxation (cf. Section 1.1). The resulting decrease in $(n_1 - n_2)$, the population excess which is

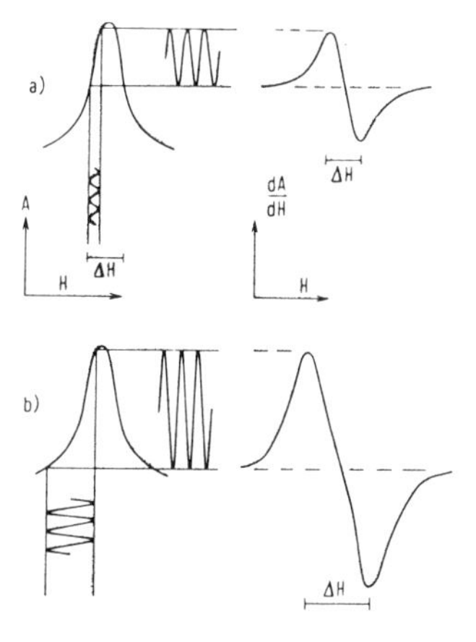

Figure 5: Modulation with an amplitude(a) smaller than half the line-width ΔH and(b) greater than the line-width ΔH. In the first case, the modulated signal has retained its original width, while in the second case it is broadened and distorted (A and H are the absorption intensity and the field strength respectively).

responsible for absorption, both weakens and distorts the hyperfines lines. To prevent saturation, the intensity of the microwave radiation must be sufficiently attenuated. The signal-to noise ratio can be improved in the simplest way by raising the modulation amplitude. However, care must again be exercised, since modulation affects the line shape, and when the modulation amplitude is approximately equal to the line-width, the resolution deteriorates. Therefore, the modulation amplitude generally should not exceed about half the line-width ΔH (see Figure 5).

Internal factors. The spectrum is generally determined by the nature of the sample, decisive influence being exerted on the resolution and/or the intensity of hyperfine lines by the concentration of radical-ions, other dissolved species, and the physical properties of the solvent. These factors affect the line-width mainly through spin-spin relaxation processes occurring between the unpaired electrons of adjacent radical ions (intermolecular) as well as between the electron spin and the nuclear spins of one and the same radical-ion (intramolecular). Their influence differs so greatly from case to case that no general statement can be made as to the optimal conditions for recording the spectra. However, these conditions can be found empirically with greater ease when the significance of these factors is known.

Concentration of radical-ions. While only the intramolecular electron-nucleus interactions are of importance in dilute solutions, intermolecular interactions between the unpaired electrons of neighbouring radicals predominate in concentrated solutions. They belong to two types, namely dipole-dipole interaction and spin exchange. Dipole-dipole interaction is analogous to the interaction between an unpaired electron and a magnetic nucleus in a radical (cf. Section 1.1). As the concentration is increased, the distance between the unpaired electrons decreases, and each electron's magnetic moment (which is much larger than that of the nuclei) becomes more stronglv operative. It may be assumed that the dipole-dipole interaction between unpaired electrons is not completely averaged out by the Brownian motion of molecules even in non-viscous solutions, and that it contributes to the broadening of lines in concentrated solutions[81].

Nevertheless, for this type of line broadening, shown in Figure 6, another mechanism seems to be more effective than the dipole-dipole interaction. This mechanism[83] is the spin exchange between two unpaired electrons, 1 and 2, belonging to two neighbouring radical-ions, i.e. $M_S^{(1)} = +\frac{1}{2}, M_S^{(2)} = -\frac{1}{2} \longrightarrow M_S^{(1)} = +\frac{1}{2}. \ M_S^{(2)} = +\frac{1}{2}$.
The probability of this exchange is clearly enhanced with decreasing intermolecular distance, i.e. with increasing concentration. Since the spin exchange reduces the lifetime of a spin state, the sharpness of the ESR lines must consequently diminish in accordance with the uncertainty relation (eq. (7)). Indeed, as the frequency of the exchange increases, the hyperfine structure of the spectrum disappears, being replaced

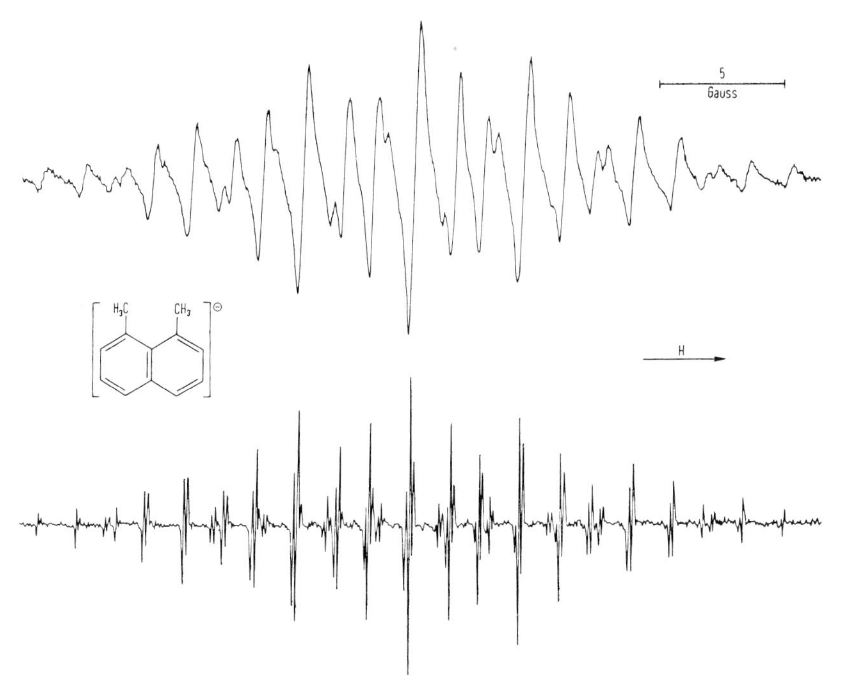

Figure 6: ESR spectra of the radical-anion of 1,8-dimethylnaphthalene at various concentrations [ca. 10^{-3} mole/lit. (top) and ca. 10^{-5} mole/lit. (bottom)] in 1,2-dimethoxyethane, at -70 °C; gegenion: $Na^{\oplus}$ [82].

by an undifferentiated and broad band. Such a band narrows at extremely high radical concentrations (exchange narrowing).

To achieve high resolution, therefore, one should use highly dilute solutions. However, a limit is set to the dilution by the sensitivity of the spectrometer, which thus indirectly affects the resolution.

Effect of co-solutes. The spectrum is influenced not only by interactions between the radical-ions, but also by interactions between the radical-ions on one hand and co-solutes (paramagnetic or diamagnetic species) on the other. The interactions of radical-ions with paramagnetic species are of the same type as the interactions between radical-ions. Figure 7 shows the effect of dissolved oxygen on the ESR spectrum of Wurster's Blue (radical-cation of N,N,N'N'-tetramethylphenylenediamine)[81,84]. One can remove the oxygen by degassing the solution under vacuum or flushing it with nitrogen, though it is best to do both.

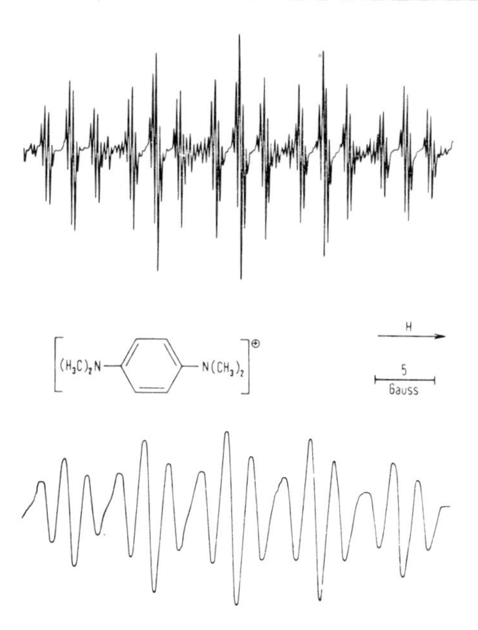

Figure 7: Central part of the ESR spectra of Wurster's Blue perchlorate in oxygen-free ethanol (top) and in solution saturated with air (bottom)[84], at + 25 °C.

The effect of diamagnetic species on the resolution is manifested when the radical-anions ($Ar^{\ominus}$) are accompanied by the corresponding neutral molecules (Ar) and/or the doubly charged anions ($Ar^{\ominus\ominus}$)[85-87,229]. These species differ from the radical-anion by only one π-electron, so their structures are very similar. What is involved is a transfer of π-electrons:

$$\begin{aligned} Ar^{\ominus} + Ar &\longrightarrow Ar + Ar^{\ominus}; \\ Ar^{\ominus} + Ar^{\ominus\ominus} &\longrightarrow Ar^{\ominus\ominus} + Ar^{\ominus} \end{aligned} \qquad (17)$$

Like spin exchange between the electrons of two neighbouring radical-ions, this transfer reduces the lifetime of a spin state and broadens the hyperfine lines. Good resolution therefore necessitates complete conversion of the compound into the radical-anion, which, however, is possible only when the starting compound does not form doubly charged anions under the conditions of the preparation; otherwise the reduction must be prolonged until the equilibrium $Ar + Ar^{\ominus\ominus} \rightleftharpoons 2\,Ar^{\ominus}$ lies as much to the right as possible (cf. Section 1.2).

One can generally neglect the interactions between radical-ions and diamagnetic species which have a completely different electronic structure, except for alkali metal cations associating with radical-anions in low-polarity solvents such as 1,2-dimethoxyethane or tetrahydrofuran (cf. detailed discussion in Appendix A.2.2). Such association is clearly proved by an additional hyperfine splitting due to alkali metal nuclei, though this splitting is observed only in the case of tight ion-pairs. The association is frequently a loose one, so that the cation continuosly changes its position relative to the radical-anion. As a result, there are small fluctuations in the coupling constants, giving rise to a broadening of the hyperfine lines. Such an association therefore affects the resolution unfavourably, and should be suppressed as much as possible. This can be done by cooling the sample, thus increasing the dielectric constant of the solvent, which has a decisive influence on the association. Figure 8 shows the effect of cooling on the spectrum obtained for the radical-anion of cycl [3,2,2,] azine[88].

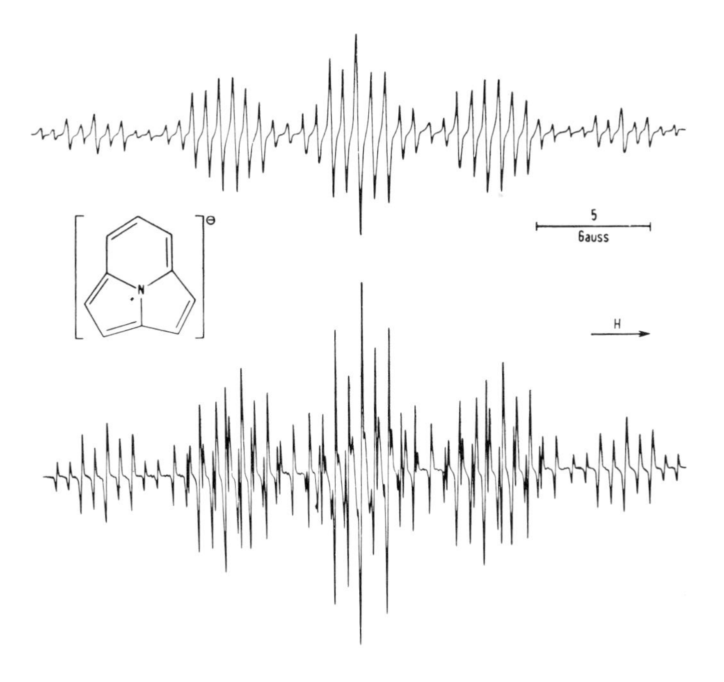

Figure 8: ESR spectra of the radical-anion of cycl[3,2,2]azine in 1,2-dimethoxyethane, at +25 °C (top) and −60 °C (bottom)[88]; gegenion: $Na^{\oplus}$.

Despite having a lower dielectric constant, 1,2-dimethoxyethane promotes the association to a smaller extent than tetrahydrofuran, presumably because of the better solvation of cations in it (cf. Appendix A2.2). Therefore, solutions in 1,2-dimethoxyethane generally give better-resolved spectra of radical-anions than solutions in tetrahydrofuran at the same temperature and concentration.

Physical properties of solvents. Properties such as the polarity and the solvating power of solvents can affect the association between the radical-anions and their gegenions, and thus the resolution of the spectrum. Furthermore, the viscosity of the solvent may influence the line-width by the following mechanism. As already mentioned in Section 1.1, the Brownian motion of the particles in the liquid does not lead to complete averaging-out of the intramolecular dipole-dipole interaction between unpaired electrons and magnetic nuclei within a radical-ion; instead, a residue remains which contributes to the line-width and which is, as expected, greater in the case of viscous liquids than non-viscous ones.

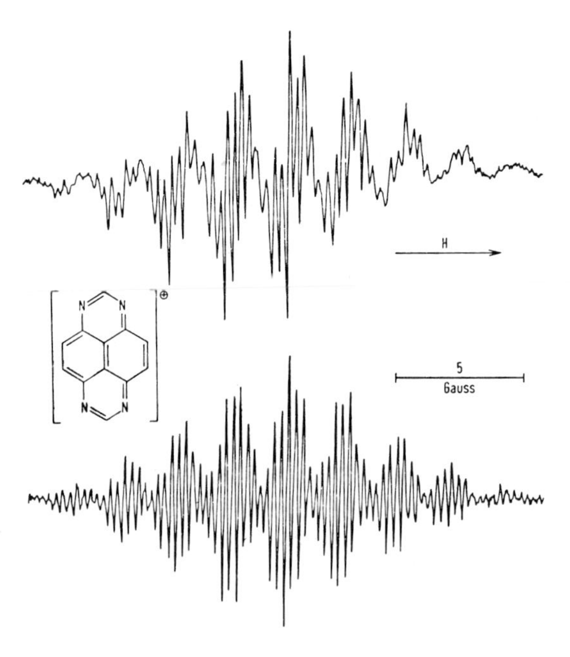

Figure 9: ESR spectrum of the radical cation of 1,3,6,8-tetraazapyrene in conc. sulphuric acid, at +25 °C (top) and +80 °C (bottom)[58].

Concentrated sulphuric acid, a rather viscous liquid, is used as a solvent to record the spectra of those radical-cations which can be prepared from some aromatic hydrocarbons in this medium (cf. Section 1.2). The lines of the resulting spectra are frequently broader than they are when less viscous solvents are used. The viscosity of liquids is known to decrease as the temperature increases, and for concentrated sulphuric acid it amounts to 0.254 and 0.0519 poise at +20 and +80 °C, respectively [89]. In view of this, it is advantageous to heat the sample, provided, of course, that the radical-cations are stable enough[90]. Figure 9 shows the effect of temperature on the resolution of the spectrum obtained for the radical-cation of 1,3,6,8-tetraazapyrene in conc. sulfuric acid[58].

1.4. Analysis of the hyperfine structure

Simple cases. Using the rule given in Section 1.1, one should be able to determine the coupling constants from the distances between the hyperfine components and the relative peak intensities provided that all or nearly all the lines of the spectrum are observed. However, this has as yet been done only in the case of a few radical-ions which exhibit relatively simple hyperfine patterns in their spectra. According to eq.

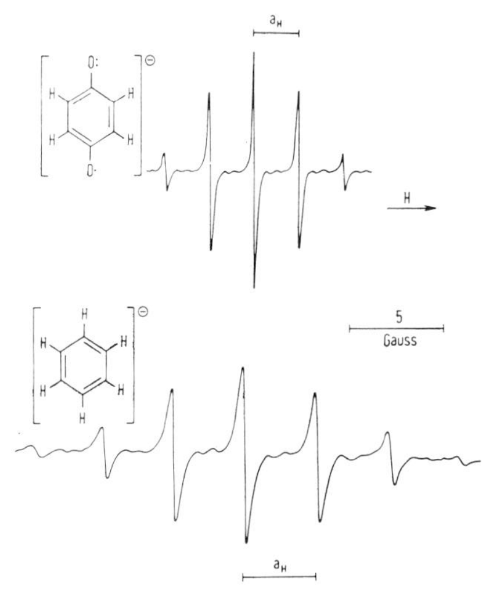

Figure 10: ESR spectra of the radical-anion of 1,4-benzoquinone (semiquinone-anion) and of benzene.
Top: in 80% alkaline ethanol at +25 °C; gegenion: $Na^{\oplus}$.
bottom: in 1,2-dimethoxyethane at −80 °C; gegenion: $K^{\oplus}$.

(16), the number of lines is determined by the number of sets of equivalent magnetic nuclei in the molecule and by the number of such nuclei in each set. It follows that the smaller radical-ions characterized by high symmetry are expected to present a simple hyperfine structure.

Figure 10 shows the spectra of two radical-ions each containing only one set of equivalent magnetic nuclei. The interaction of the unpaired electron with the four magnetic protons in the 1,4-benzosemiquinone-anion[25a)] gives rise to five equidistant lines, the analogous interaction with the six protons in the radical-anion of benzene[30,36)] giving rise to seven equidistant lines. Thus, as expected, the interaction with n equivalent protons leads to n + 1 hyperfine components, whose intensity distribution is governed by the binomial coefficients of the nth order. The relative intensities of these lines in the spectra of 1,4-benzosemiquinone-anion and

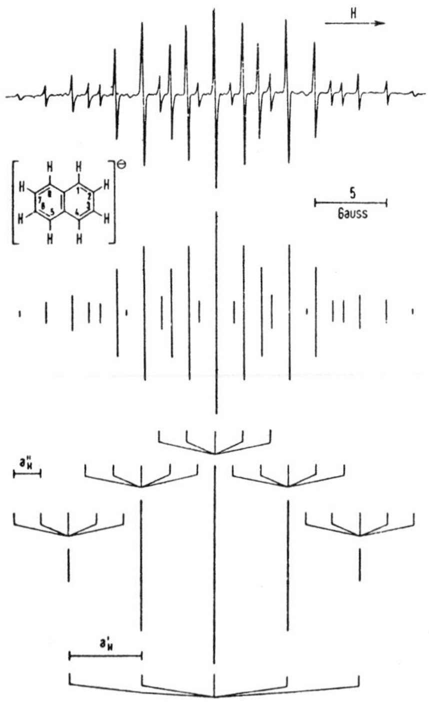

Figure 11: ESR spectrum of the radical-anion of naphthalene (in 1,2-dimethoxyethane, at +25°C; gegenion: $Na^{\oplus}$) and the reconstructed hyperfine structure (stick diagram, drawn in stages from the bottom upwards).

of benzene radical-anion are respectively 1:4:6:4:1 and 1:6:15:20:15:6:1 (cf. Pascal's triangle in Section 1.1).

The distance between two adjacent lines gives the coupling constant a_H, the value of which is 2.37 gauss for the four equivalent protons in the 1,4-benzonesemiquinone-anion and 3.75 gauss for the six equivalent protons in the radical-anion of benzene. When the radical-ion contains several sets of equivalent magnetic nuclei, it is useful to reconstruct the hyperfine structure with the aid of the empirical coupling constants. The analysis is then correct if the reconstructed spectrum agrees with the recorded one. This point will now be illustrated on two simple examples, first on the radical-anion of naphthalene (see Figure 11), and then on that of 1,4,5,8-tetraazanaphthalene (see Figure 12).

The radical-anion of naphthalene [24,44,91] contains two sets of equivalent magnetic nuclei: one set comprises four protons in positions 1,4,5 and 8; and the other set comprises the remaining four protons in positions 2,3,6 and 7. Each set must give rise to five equidistant hyperfine lines with an intensity distribution of 1:4:6:4:1. Provided that no accidental degeneracy is involved (see below), the inter-line distance will be different for the two sets and will correspond to the coupling constants a_H' and a_H''. When the spectrum has been recorded, the two coupling constants can be determined, e.g. by measuring the distances between the most intense central line and each of the two next highest peaks on one side of the spectrum: $a_H' = 4.95$ and $a_H'' = 1.83$ gauss [82]. The spectrum is then reconstructed by drawing five lines with an intensity ratio of 1:4:6:4:1 and with an inter-line separation of 4.95 gauss, and then by splitting each of these into five lines again with an intensity distribution of 1:4:6:4:1, but now with an inter-line separation of only 1.83 gauss. This last operation amounts to drawing both a 4/6 th-intensity and a 1/6 th-intensity line on both sides of each of the first five lines. One thus obtains a hyperfine structure consisting of 5 x 5 = 25 components. Comparison shows that this reconstructed structure fully agrees with the recorded spectrum (Figure 11).

However, it is not possible to tell from the spectrum which set of protons has a coupling constant of 4.95 gauss and which 1.83 gauss. This is because both sets give a fivefold splitting with an intensity ratio of 1:4:6:4:1. In fact, such an indeterminacy of assignment exists whenever a radical-ion contains more than one set of equivalent magnetic nuclei, each set giving rise to the same number of equidistant lines with identical intensity distribution, differing only in the inter-line separation. The methods permitting an assignment to be made in such cases will be discussed at the end of this section.

The second example, the radical-anion of 1,4,5,8-tetraazanaphthalene[92], differs from the first in the fact that the two coupling constands a_H and a_N can be unambiguously assigned to two sets of four nuclei (four protons and four nitrogens), since these sets give rise to dissimilar hyperfine structures (cf. intensity schemes in Section 1.1). The recorded and the reconstructed ESR spectra of this radical-anion are shown in Figure 12.

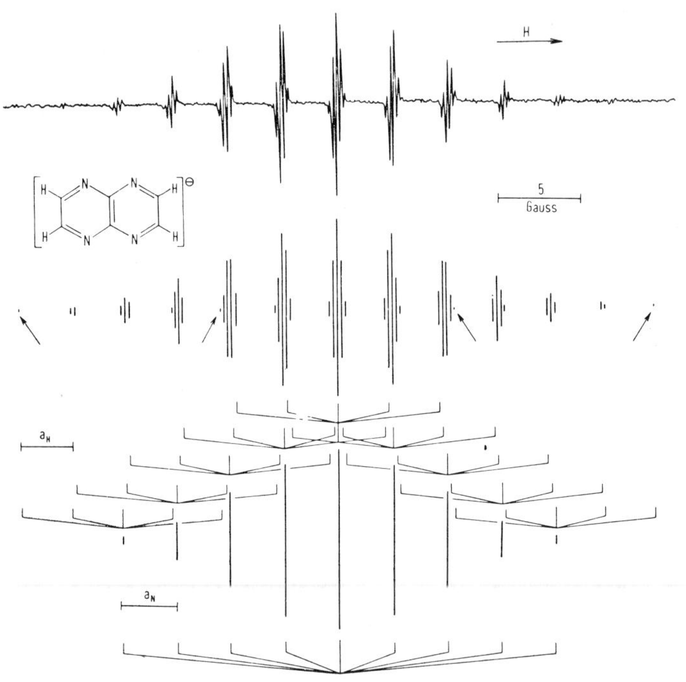

Figure 12: ESR spectrum of the radical-anion of 1,4,5,8-tetraazanaphthalene (in N,N-dimethylformamide, at +25°C; gegenion: tetraethylammonium ⊕) [92] and the reconstructed hyperfine structure (stick diagram, drawn in stages from the bottom upwards).

Coupling with the four equivalent ^{14}N nuclei results in nine equidistant lines (^{14}N has a spin of 1) with an inter-line separation of $a_N = 3.37$ gauss and an intensity distribution of 1:4:10:16:19:16:10:4:1. Coupling with the four equivalent protons, furthermore, leads to a fivefold splitting of each of the above nine lines, the inter-line separation being here $a_H = 3.14$ gauss and the intensity distribution of 1:4:6:4:1. The observed grouping into sets of equidistant lines is due to the fact that the difference between the two coupling constants a_N and a_H (which gives the splitting inside the groups) is much smaller than the constants a_H and a_N themselves (which determine the distance between adjacent groups). Were the two coupling constants identical, the observed groups would become single lines. The recorded spectrum in Fig. 12 contains 41 of the expected 9 x 5 = 45 lines, but the four weakest missing lines (marked by arrows) can also be identified by the solutions with a higher radical concentration.

Complex cases. The more numerous the lines, and the more frequently they overlap, the more difficult the analysis becomes. A relatively complex hyperfine structure with many superpositions of lines is exhibited by the ESR spectrum of the radical-anion of 2,3-dimethylnaphthalene reproduced in Figure 13[82].

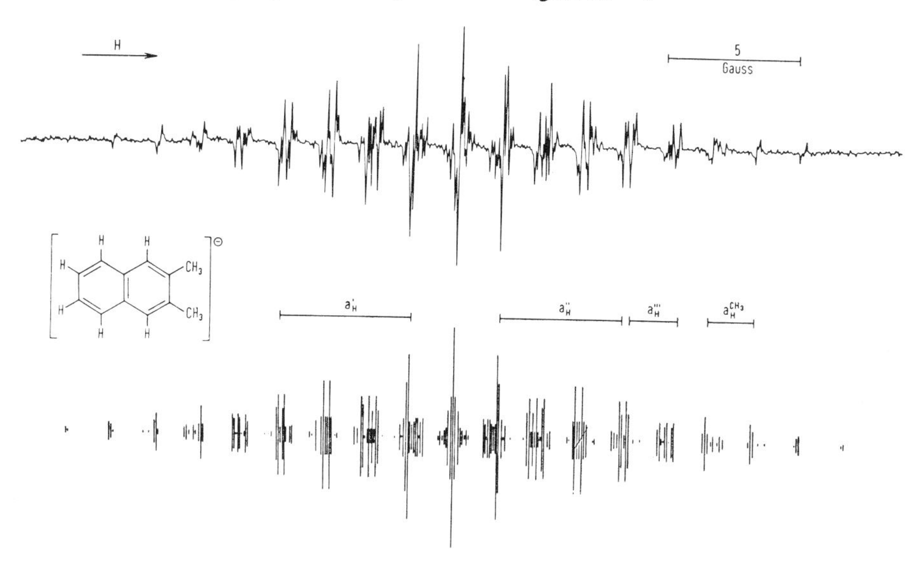

Figure 13: ESR spectrum of the radical-anion of 2,3-dimethylnaphthalene (in 1,2-dimethoxyethane, at –70 °C; gegenion: $Na^{\oplus}$)[82] and the reconstructed hyperfine structure (stick diagram).

Although only about half of the expected 3^3 x 7 = 189 lines are fully resolved, the spectrum can be analysed without much difficulty on the basis of the easily identifiable strongest lines in the middle. However, in many other cases where extensive superposition in the middle does not permit unambiguous determination of the coupling constants, greater attention must be paid to the two ends of the spectrum. Since the end lines are generally rather weak, it is advisable to record the spectrum in these edge regions at a better signal-to-noise ratio, achieved by a higher radical concentration and a larger modulation amplitude.

It has become common practice to reconstruct spectra not merely as schemes of separate hyperfine lines (stick diagrams), but as continuous curves simulating recorded empirical spectra [93]. This is done by computers programmed on the basis of the parameters determining the hyperfine structure (number of lines, intensity distribution, and coupling constants) and the appearance of the lines, i.e. the shape and the width of the components (Gaussian or Lorentzian curves). This process is particularly suitable when the empirical spectrum is complex and insufficiently resolved, thus exhibiting superimposed lines. The reconstruction of the lines by a stick diagram is then not a reliable basis of comparison for the analysis of the recorded

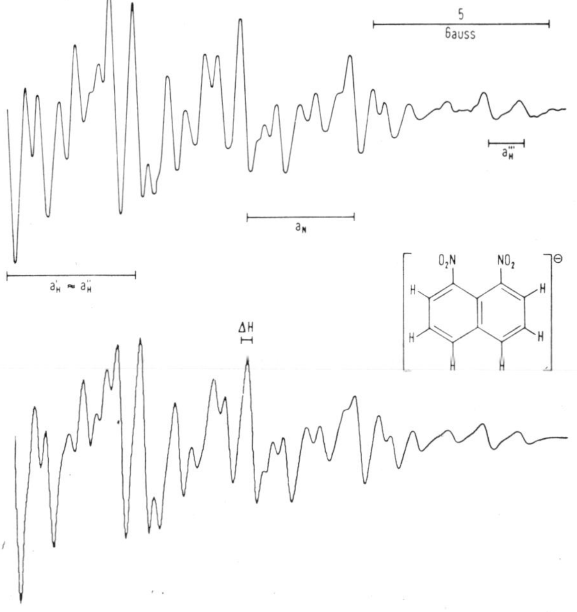

Figure 14: ESR spectrum of the radical-anion of 1,8-dinitronaphthalene (low-field half) [94].
Top: recorded in N,N-dimethylformamide, at +25 °C; gegenion: tetrethylammonium$^{\oplus}$
bottom: calculated on the basis of the following values:
a_N = 3.07 gauss, as coupling constant of two equivalent ^{14}N nuclei;
a_H' = 3.73 gauss, as coupling constant of four equivalent protons (accidental degeneracy);
a_H'' = 1.00 gauss, as coupling constant of two equivalent protons;
ΔH = 0.31 gauss, as the width of hyperfine lines (Lorentzian curves).

spectrum. The observed and the computer-simulated spectra of the radical-anion of 1,8-dinitronaphthalene are compared in Figure 14[94].

Accidental degeneracy. Nearly exact superposition of two or more lines often leads to a single, practically unbroadened line whose intensity is roughly the sum of the intensities of the telescoped lines. This is observed when the coupling constants are connected by relationships that can be approximately expressed by simple integers. One speaks then of "accidental degeneracy", meaning a degeneracy that is manifested only at the given experimental conditions and can be made to disappear by improvement of the resolution. In other words, in contrast to intrinsic degeneracy, accidental degeneracy arises from the imperfection of the conditions of the measurement and not from molecular symmetry.

Whenever fewer lines are found than expected and they show an anomalous intensity ratio, accidental degeneracy should be suspected. In such cases, the relationship between the coupling constants must be estimated and then verified by reconstruction of the hyperfine structure. This point will now be illustrated by the spectrum of the semiquinone-anion obtained from trans-15,16-dimethyldihydropyrene-2,7-quinone[54]. (see Figure 15).

The five groups of lines exhibiting an intensity ratio of 1:4:6:4:1 are due to coupling with four equivalent ring protons, characterized by the largest constant $a_H' =$ 1.66 gauss. Each group is expected to comprise 5 x 7 = 35 lines, the fivefold splitting originating from the second set of four equivalent ring protons (coupling constant a_H''), and the sevenfold splitting arising from the six equivalent methyl protons (coupling constant $a_H^{CH_3}$). Contrary to expectations, however, each group consists of only 15 equidistant lines in an intensity ratio that does not tally with any of the rows in Pascal's triangle (cf. Section 1.1). The assumption that $a_H'' = 2a_H^{CH_3} = 0.224$ gauss gives 15 lines in each group, characterized by the following distribution of intensities:

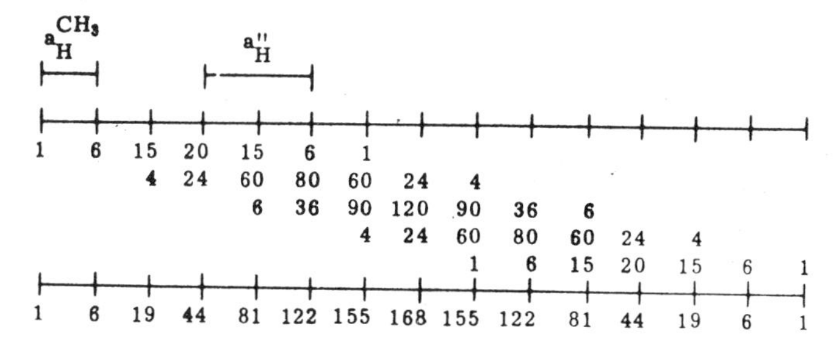

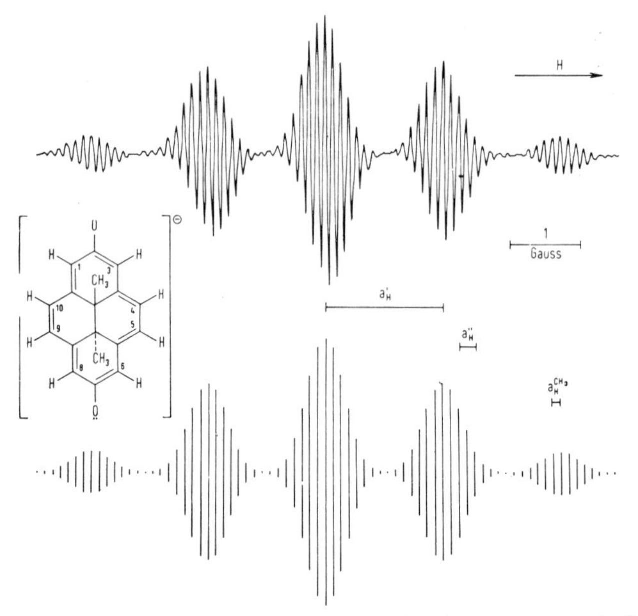

Figure 15: ESR spectrum of the trans-15,16-dimethyldihydropyrene-2,7-semiquinone-anion(in 80% alkaline ethanol, at +25 °C; gegenion: $Na^{\oplus}$)[54] and the reconstructed hyperfine structure (stick diagram).

Comparison between the recorded and the reconstructed spectra in Fig. 15 shows that this assumption leads to perfect agreement between expected and observed intensities. The assignment of the two coupling constants a_H' and a_H'' to the two sets of four equivalent ring protons (first set: positions 1, 3, 6 and 8; second set: positions 4, 5, 9 and 10) is once more based not on the spectrum, but on a method to be described below.

Assignment of the coupling constants. As mentioned before, coupling constants cannot be assigned unambiguously to sets of equivalent nuclei on the basis of the spectrum in cases where these sets do not differ in the resulting number of lines and in their intensity distribution. This happens when the radical-ion contains sets each consisting of the same number of equivalent nuclei, and when these nuclei have the same spin quantum number I.

A frequent problem in the analysis of the spectra of aromatic radical-ions is the assignment of coupling constants to equally large sets of equivalent ring protons.

The most reliable method of solving this problem is based on specific deuteration. While the electron structure of the radical-ion is little affected*) by substitution of a deuteron for a proton, the magnetic properties of the nucleus in question are drastically changed by such a replacement (cf. Table 1). Since the spin quantum number I of the deuteron is 1 and not $\frac{1}{2}$, n equivalent deuterons give the same number

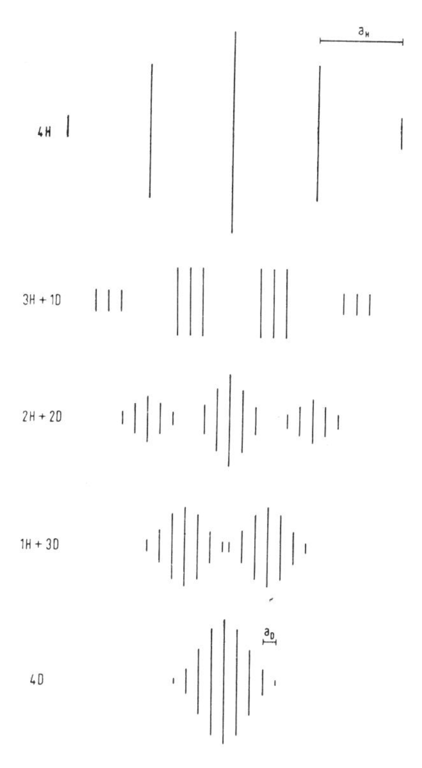

Figure 16: Changes in the hyperfine structure accompanying the stepwise replacement of four equivalent protons by deuterons (the total intensity is kept the same throughout in this schematic representation).

---

*) One exception is represented by radical-ions having a degenerate ground state, in which specific deuteration can lift the degeneracy and thus cause a spin redistribution (cf. Section 2.5). For radical-ions in a non-degenerate ground state the perturbation brought about by deuteration is usually small. It can be estimated from the changes of the coupling constants of the protons in non-deuterated positions. In the case of the naphthalene radical-anion, for example, these changes are 0.04 gauss at most[328].

of lines with the same intensity distribution as n equivalent ^{14}N nuclei, and not as n equivalent protons (cf. Figures 2 and 3, and intensity schemes in Section 1.1). Besides, the replacement of a proton (H) by a deuteron (D) reduces the inter-line distance by a factor of $g_N(D)/g_N(H) = 0.8574/5.5854 = 0.1535$, so that the overall range of the spectrum diminishes despite an increase in the number of lines.

Figure 16 shows schematically the hyperfine structure resulting from the progressive replacement of four equivalent protons by deuterons. Such a replacement is seen to affect visibly only the hyperfine structure due to the nucleus in question,*) and so comparison of the spectra before and after shows which coupling constant is changed by the deuteration. Then, provided that the position of the deuteron in the molecule is known, the coupling constants can be assigned unambiguously to the correct sets of equivalent protons. This technique has been used with the radical-anion of naphthalene which contains two sets of four equivalent protons, and revealed that the larger coupling constant $a_H' = 4.95$ gauss belongs to the protons in positions 1, 4, 5 and 8 the smaller value $a_H'' = 1.83$ gauss belonging to the protons in positions 2,3,6 and 7 [95,96].

The type of comparison involved in this assignment is shown in Figure 17 in which the spectra of the radical-anions of naphthalene and its 1,4,5,8-tetradeuterio-derivative are reproduced [97].

This technique has been rather seldom used, despite being elegant and reliable, because deuterio-derivatives of aromatic compounds are often not readily accessible. The preparation of such derivatives is generally based on H/D exchange in a deuteron-rich medium [98], or on the replacement of reactive substituents (e.g. halogens) by deuterium.

The difficulties involved gave an impetus to the development of alternative methods of assignment of the coupling constants to sets of equivalent protons. In one of these methods [99], use is made of the broadening of the proton resonance signals by the addition of a small amount of a radical-anion $Ar^{\ominus}$ to a solution of the neutral parent compound Ar. The broadening is proportional to the coupling constant $a_{H\mu}$ of the radical-anion's proton $H\mu$ in question. The method is suitable for compounds whose NMR spectra can be analysed with certainty.

When experimental methods fail or become too involved, chemists resort to analogy, as in other spectroscopic work. It is reasonable to expect that the coupling constants of ring protons will be rather similar when the radical-ions have similar structures. This is the basic assumption that permits the assignment of the coupling constants in a radical-ion by the use of the necessary experimental data for a structurally related

*) See footnote on the preceding page.

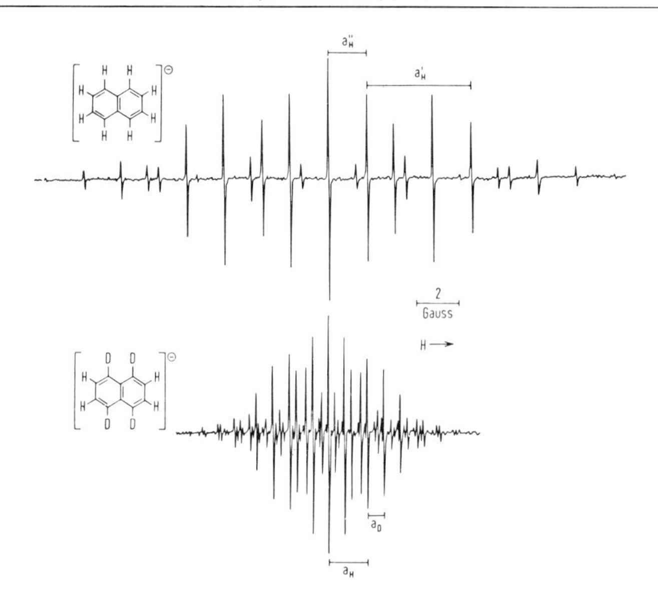

Figure 17: ESR spectra of the radical-anions of naphthalene and its 1,4,5,8-tetradeuterio-derivative in 1,2-dimethoxyethane, at $-70\ ^{\circ}C$; gegenion: $Na^{\oplus}$. $a_D \approx 0.1535\ a_{H'}$.[97]

known radical-ion (radical-ions having a degenerate or almost degenerate ground state again require a careful approach; cf. Section 2.5). The more numerous the experimental data and the smaller the structural differences, the more reliable will be the assignment. The radical-anion of 1,8-dimethylnaphthalene, whose spectrum has already been shown in a different context (see Figure 6), may be taken as an example. The experimental values[82] for the coupling constants of the three pairs of equivalent ring protons are as follows: $a_{H'} = 4.71$ gauss and $a_{H''} \approx a_{H'''} = 1.70$ gauss (accidental degeneracy), but these constants cannot be assigned from the spectrum alone. However, utilizing the results of analogous investigations[50,51], one can assume that the introduction of two methyl groups into the radical-anion of naphthalene does not modify fundamentally the latter's ring proton coupling constants, so that in methyl derivatives, too, the greater coupling constant should belong to the ring protons in positions 1,4,5 and 8. Of such protons, there are only two in 1,8-dimethylnaphthalene, namely those in positions 4 and 5.

H 1,83 ; 4,95 (naphthalene) — H_3C, CH_3; H 1,70 ; H 1,70 ; 4,71 (1,8-dimethylnaphthalene)

The superposition model[100], which correlates the experimental data for variously substituted derivatives of a radical-ion, is also useful for comparative purposes. The most frequently used method for assigning coupling constants to sets of equivalent ring protons is based on a linear relationship between the coupling constant of the ring proton $a_{H\mu}$ and the theoretical quantity ρ_μ^π (π-spin population at the centre μ). The reliability of the method depends primarily on the accuracy of calculating ρ_μ^π. It will emerge from a discussion of the MO theory of aromatic radical-ions (Section 1.5 and 1.6) that ρ_μ^π is calculated by approximation, and so the assignment in this case is certain only when the coupling constants differ quite appreciably. This is so for the 2,7-semiquinone of 15,16-dimethyldihydropyrene (cf. Fig. 15), where a_H = 1.66 and a_H'' = 0.224 gauss. Calculations give much higher values ρ_μ^π for the centres μ = 4,5,9 and 10 than for μ = 1,3,6 and 8, so that the larger coupling constant can be assigned to the four equivalent protons at the first set of centres and the smaller to the four protons at the second set[54].

The problem of assigning coupling constants to ^{14}N nuclei and to protons in alkyl substituents arises much less often, and is handled by the same methods, namely by $^{14}N/^{15}N$ and H/D exchange, respectively, use of analogy, and MO considerations. To assign coupling constants $a_{N\mu}$ to ^{14}N nuclei, use can also be made of line-width investigations, such as those carried out in connection with ^{13}C satellites (cf. Appendix A 2.1) in the ESR spectrum of the anthracene radical-anion[101].

Calibration of the spectrum. The presentation of the coupling constants in gauss requires that abscissa of the spectrum must be calibrated in this unit. The field strength H is roughly indicated on certain commercial ESR spectrometers (e.g. Varian Fieldial), but an unknown spectrum can be calibrated more accurately against a known one if both are recorded simultaneously with the aid of a dual sample cavity and a two-channel recorder. A frequently used reference substance is the dianion of nitrosobissulfonic acid $ON(SO_3)_2^{\ominus\ominus}$ whose spectrum, due to coupling with the ^{14}N nucleus of the nitroso group, exhibits three equidistant lines having the same intensity and a separation of 13.0 ± 0.1 gauss[102]. The determination of coupling constants on this basis involves a maximum error of about + 1%. A still more accurate measurement of the field strength is based on proton magnetic resonance. In this technique the field strength H is found from the proton resonance frequency ν with the aid of the formula $\nu = \gamma_N$ H, where γ_N = 4.258 kHz · gauss^{-1} (this formula is analogous to eq. (5) for the electron).

1.5. Theoretical interpretation

Ring protons. As mentioned in Section 1.1, the coupling constant of a proton a_H is directly proportional to the spin density $\rho'(0)$ at this proton.

$$a_H = K_H \rho'(0) \tag{14}$$

The spin density $\rho'(\mathbf{r})$ is the difference between two electron densities $q'_\alpha(\mathbf{r})$ and $q'_\beta(\mathbf{r})$ which are made up of electrons differing in their spin quantum number M_S, i.e.

$$\rho'(\mathbf{r}) = q'_\alpha(\mathbf{r}) - q'_\beta(\mathbf{r}), \tag{18}$$

where $q'_\alpha(\mathbf{r})$ is the density of electrons with $M_S = +\frac{1}{2}$ (i.e. α or ↑ spins), $q'_\beta(\mathbf{r})$ is the density of electrons with $M_S = -\frac{1}{2}$ (i.e. β or ↓ spins), and $\mathbf{r}$ is a space vector defining their positions ($\mathbf{r} = 0$ at the proton).

When electrons of different spins occupy the same orbitals in pairs, the contributions to q'_α and q'_β cancel each other, giving a zero resultant spin density. In a diamagnetic substance, therefore, the spin density $\rho'(\mathbf{r})$ is everywhere zero. In a paramagnetic species, on the other hand, the unpaired electron is in a singly occupied orbital, and the spin density does not vanish everywhere. By convention, the unpaired electron is given a spin quantum number of $M_S = +\frac{1}{2}$, and therefore $\rho'(\mathbf{r})$ should assume only positive values. As will be seen later, however, negative spin densities are also possible at certain sites in paramagnetic multi-electronic systems.

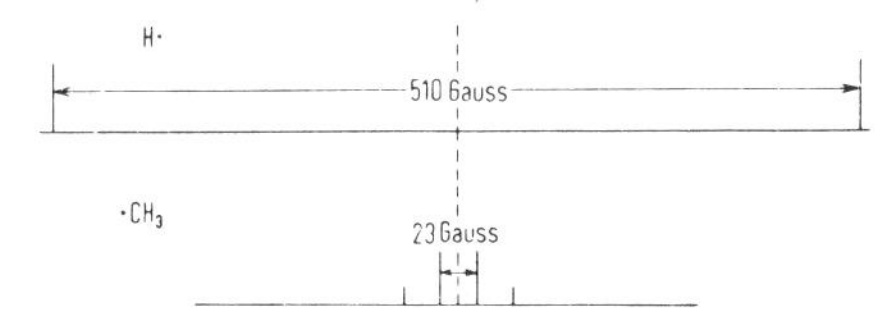

Figure 18: Schematic hyperfine structures of atomic hydrogen and the methyl radical.

The hyperfine structure of two simple paramagnetic species, the hydrogen atom ·H and the methyl radical $\cdot CH_3$, are shown schematically in Figure 18. In the hydrogen atom with its single electron in the ls-orbital, the spin density $\rho'(\mathbf{r})$ is everywhere positive and equal to the electron density, i.e. to the square of the ls-wave-function. The value $1/\pi r_o^3$ of this square at the site of the proton gives the constant a_H on substitution for $\rho'(0)$ in eq. (14). Using

$$K_H = 2.3626 \times 10^{-22} \text{gauss. cm}^3 \text{ (cf. eq. (15)), and } r_o = 0.529 \times 10^{-8} \text{ cm},$$

one obtains $a_H = +508$ gauss, in excellent agreement with the experimental values[16)]

of 504-512 gauss (it is only the magnitude of coupling constants that can be determined from the isotropic hyperfine structure of the ESR-spectra).
In the planar methyl radical, the unpaired electron occupies a $2p_z$-orbital of the sp^2-hybridized carbon atom. The nodal plane of this orbital (i.e. the plane on which the wave-function of the orbital has a zero value) coincides in the $\cdot CH_3$ radical with the plane that traverses the carbon and the three hydrogens. It might be expected that the spin density $\rho'(\mathbf{r})$ in the methyl radical should equal the square of the $2p_z$-function for the corresponding singly occupied orbital, because neither the two paired electrons in the non-bonding ls-orbital of carbon, nor the six electrons in the doubly occupied σ-orbitals of the three C-H bonds should make any contribution to $\rho'(\mathbf{r})$. Then, the spin density must be zero in the nodal plane of the $2p_z$-orbital and have a positive value outside it. The spin density at the three protons, $\rho'(0)$, lying in the nodal plane of the $2p_z$-orbital would thus become zero, so that no hyperfine structure should appear for the $\cdot CH_3$ radical. In reality, however, the spectrum exhibits four equidistant lines having an intensity distribution of 1:3:3:1, which is expected when an unpaired electron interacts with three equivalent protons (cf. Figure 18). The coupling constant a_H is about 23 gauss[16], admittedly a low value in comparison with 510 gauss for $\cdot H$, but certainly far from zero. It must therefore be concluded that in the case of the methyl radical the assumption that $\rho'(\mathbf{r})$ can be expressed simply as the square of the $2p_z$-function does not agree with the experimental result. A similar disagreement is encountered in the hyperfine structure of ring protons in aromatic radical-ions. Not only the sp^2-hybridized carbon atoms, but also the hydrogens in these radicals lie in the nodal plane of the π-orbitals, so that, equating the spin density $\rho'(\mathbf{r})$ to the square of the function of the singly occupied π-orbital, one should find no hyperfine structure due to the ring protons. The disagreement between theory and experiment results from neglecting the interaction between the unpaired spin of $2p_z$-(or the π-electron) and the paired spins of the σ-electrons[103-107]. It will be shown in Appendix A 1.1 how the magnitude of this interaction can be estimated with the aid of the MO theory[104]. The kind of σ-$2p_z$ or σ-π interaction involved is illustrated below by a simple scheme.

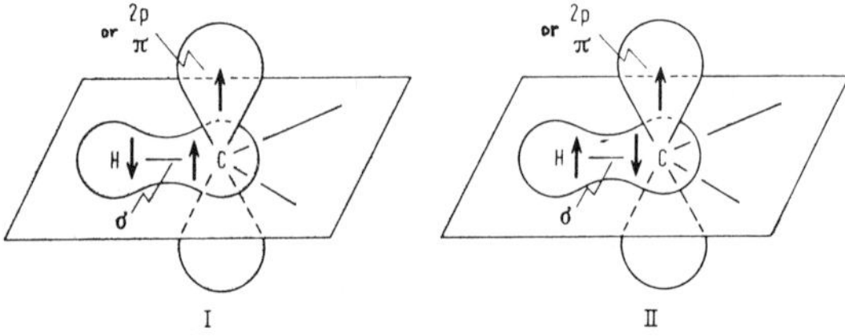

Figure 19: C-H section of the methyl radical or an aromatic radical-ion. (The σ-electron near the carbon has the same spin quantum number M_S as the unpaired $2p_z$ or π-electron in arrangement I, but a different M_S-value in arrangement II).

Figure 19 shows two spin arrangements that are possible for a C-H section of the methyl radical or an aromatic radical-ion when one of the two σ-electrons is close to the carbon and the other to the hydrogen. In case I, the σ-electron near the carbon has the same M_S-value as the unpaired $2p_z$- or π-electron, i.e. $M_S(\sigma) = M_S$ ($2p_z$ or π) $= +\frac{1}{2}$ (↑). In case II, on the other hand, these electrons differ in their spin quantum number, i.e. $M_S(\sigma) = -\frac{1}{2}$ (↓) and M_S ($2p_z$ or π) $= +\frac{1}{2}$ (↑). Arrangement I with two parallel spins represents the component $M_S^{tot} = 1$ of the triplet state of carbon ($S^{tot} = 1$), while arrangement II with opposite spins corresponds to the singlet state of carbon ($M_S^{tot} = 0$; $S^{tot} = 0$) (cf. Section 1.1). Arrangement I will be somewhat more frequent than arrangement II, because, in the case of equal occupancy of orbitals and differing spin configurations in an atom, the configuration that gives rise to the highest spin multiplicity ($2S^{tot} + 1$) is energetically favoured (Hund's rule)[17]. Thus the carbon's triplet configuration (I), where $2S^{tot} + 1 = 3$, will enjoy a slight preference over the singlet configuration (II), where $2S^{tot} + 1 = 1$, and there will be a small excess of σ-electron density with $M_S = +\frac{1}{2}$ (q'_α) at the carbon. Moreover, since the spins of the σ-electron involved in a C-H bond are oppositely oriented, there will be an equally small excess of σ-electron density with $M_S = -\frac{1}{2}$ (q'_β) at the hydrogen. This means that a small spin density $\rho'(0)$ has been polarized at the proton, with a sign opposite to that of the spin density in the $2p_z$-orbital of the methyl radical or the delocalized π-orbital of an aromatic radical-ion[107]. If the second spin density is positive, the first must be negative. The mechanism responsible for the appearance of a spin density in a σ-orbital is called σ-$2p_z$ or σ-π spin polarization. Quantitative calculations show that the magnitude of this polarized spin density is sufficiently high to give rise to a coupling constant of about 23 gauss, which has been observed for the protons of the $\cdot CH_3$ radical[106].

Rather similar values have been found for the overall range A of the spectra of unsubstituted aromatic radical-ions, i.e. for the distance between the two terminal hyperfine lines, A being given as the sum

$$A = \sum_{\mu} |a_{H\mu}| , \qquad (19)$$

where the terms $a_{H\mu}$ represent the coupling constants of the protons attached to the carbon atoms μ of the hydrocarbon. Thus, the value of A is 22.5 and 27.12 gauss for the radical-anions of benzene (Figure 10) and naphthalene (Figure 11), respectively. These values arise in the following manner:

benzene: $a_{H1} = a_{H2} = \ldots = a_{H6} = 3.75$ gauss, and $6 \times 3.75 = 22.5$ gauss;
naphthalene: $a_{H1} = a_{H4} = a_{H5} = a_{H8} = 4.95$ gauss,
$a_{H2} = a_{H3} = a_{H6} = a_{H7} = 1.83$ gauss, and
$(4 \times 4.95) + (4 \times 1.83) = 27.12$ gauss.

The reason why the individual coupling constants $a_{H\mu}$ of the ring protons are generally much smaller than 23 gauss is of course that, while the $2p_z$-electron in the methyl radical is localized on one carbon atom, the π-electron in aromatic radical-ions can delocalize over several sp^2-hybridized atoms. It seems obvious that there must be a close connection between the distribution of the unpaired π-electron over individual carbon $C\mu$ and the magnitude of the polarized spin density on the adjacent ring protons $H\mu$.

C_μ—H_μ

According to McConnell's relationship[108] (eq. (20)), the coupling constant $a_{H\mu}$ of a ring proton attached to a carbon atom $C\mu$ is proportional to the π-spin population ρ_μ^π at this carbon (the concept of spin population will be explained later):

$$a_{H\mu} = Q_{CH}\,\rho_\mu^\pi\,, \tag{20}$$

where Q_{CH} is a parameter whose absolute value lies between 20 and 30 gauss. According to the MO theory, this parameter has a negative sign (cf. Appendix A.1.1), which agrees with the notion that the polarized spin density in the σ-orbital at the ring proton $H\mu$ has a sign opposite to that in the π-orbital of the adjacent carbon atom $C\mu$[107]. The negative sign of Q_{CH} has been also confirmed by experimental ESR work on crystals[310,311].

Since eq. (20) connects the observed values $a_{H\mu}$ with the computable quantity ρ_μ^π in a simple manner, it constitutes a bridge between the experimental and the theoretical branches of the ESR spectroscopy of aromatic radical-ions.

The concept of spin population. Instead of the spin density

$$\rho'(\mathbf{r}) = q'_\alpha(\mathbf{r}) - q'_\beta(\mathbf{r})\,, \tag{18}$$

it is more convenient to use the "integrated" spin density or "spin population" ρ for the interpretation of the ESR hyperfine structure of aromatic radical-ions. The spin population ρ, like the charge q, is a dimensionless quantity, whereas the spin density ρ', like the electron density q', has the dimensions of cm^{-3} (in the literature, generally both ρ and ρ' are called spin density).

The symbol ρ_μ^φ denotes a spin population obtained by integrating the spin density $\rho'(\mathbf{r})$ over the φ-type atomic orbitals (AO's) centred on atom μ. For example, ρ_H^{1s} is the spin population in the 1s-orbital of the hydrogen atom, and its value is 1; ρ_C^{2p} refers to the spin population in the $2p_z$-orbital at the carbon centre of the methyl radical, and its value is similarly 1. For π-type molecular orbitals (MO's), which are delocalized over several sp^2-hybridized centres μ and which are constructed as linear combinations of $2p_z$-AO's (ϕ_μ), ρ_μ^π is defined as the contribution of the involved AO's ϕ_μ to the total π-spin population (cf. Section 1.6). Summation of the contributions ρ_μ^π over all centres μ in the system must give the total spin population, i.e. they must add up to 1:

$$\sum_\mu \rho_\mu^\pi = 1 . \tag{21}$$

In the simple LCAO model, in which no account is taken of the interaction between the unpaired π-electron spin and the spins of the paired π-electrons, ρ_μ^π is given by the square $c_{j\mu}^2$ of that coefficient which stands for the centre μ and the singly occupied π-orbital ψ_j. In this approximation, ρ_μ^π always has a positive sign. However, when the above-mentioned interaction (π-π spin polarization) is taken into account, as in some refined MO methods, ρ_μ^π values can also appear with a negative sign. The π-π spin polarization, whose mechanism is fully analogous to that of the σ-π spin polarization, will be discussed in greater detail in Section 1.6 and Appendix A 1.2.

If eq. (21) is to be valid, the appearance of negative ρ_μ^π values must be compensated by a corresponding increase in the positive π-spin population, as a result of which the sum of the absolute magnitudes of the terms ρ_μ^π often exceeds 1, i.e.

$$\sum_\mu |\rho_\mu^\pi| \geq 1 . \tag{22}$$

Since the proportionality constant Q_{CH} in eq. (20) has a negative sign, the coupling constant $a_{H\mu}$ of a ring proton must be negative when the corresponding spin population ρ_μ^π is positive, and positive if the latter is negative. Positive spin populations ρ_μ^π are greater in magnitude and more frequent in occurence than negative ones, and therefore most coupling constants $a_{H\mu}$, and particularly those whose absolute values are large, carry a negative sign.

The isotropic hyperfine structure provides no information as to the sign of the coupling constants; nor generally do the line-widths for the coupling constants $a_{H\mu}$ of the ring protons, though they frequently do in the case of $a_{C\mu}$ and $a_{N\mu}$ of ^{13}C and ^{14}N nuclei[111,117]. Furthermore, proton resonance spectroscopy of radicals in solutions[275,276] – which was used to determine the sign of the coupling constants of protons in alkyl substituents – has rarely found application in this field[231,312]. A promising method for determining the sign of the ring proton coupling constants is based on the ESR spectra recorded for solutions of radicals in liquid crystals[295], but it is mainly limited to neutral radicals owing to the small number of radical-ions stable in suitable solvents[296]. The results obtained so far agree with the theoretical predictions. For radical-ions with only positive spin populations ρ_μ^π, the ove range A of the spectrum must be equal to or smaller than $|Q_{CH}|$, according to whether all or only some carbon centres μ carry ring protons. $|Q_{CH}|$ is therefore frequently identified as the spectral range A = 22.5 gauss measured for the radical-anion of benzene, since all six of the carbon centres μ carry protons in this species, and the D_{6h} symmetry of benzene precludes negative spin populations.

On the other hand, A may be greater than $|Q_{CH}|$ in accordance with

$$A = \sum_\mu |a_{H\mu}| = |Q_{CH}| \sum_\mu |\rho_\mu^\pi| , \tag{23}$$

if $\sum_\mu |\rho_\mu^\pi| > 1$, i.e. when negative ρ_μ^π-values are also involved. In fact, a large overall range (30 - 40 gauss) has been observed for radical-ions of those unsubstituted hydrocarbons for which theory predicts considerable negative spin populations.

^{13}C and ^{14}N nuclei in aromatic rings. While ^{14}N is by far the most abundant nitrogen isotope (and except, of course, for the proton, the most important magnetic nucleus in organic molecules), the natural abundance of the magnetic isotope ^{13}C of carbon is only 1.1% (cf. Table 1). This has been sufficient in favourable cases to determine the coupling constant $a_{C\mu}$ of ^{13}C nuclei in aromatic radical-ions, but in other cases, the ^{13}C content must be enriched (cf. Appendix A 2.1).

The ^{13}C and ^{14}N nuclei of sp^2-hybridized atoms lie in the nodal plane of the π-orbitals. Therefore, the observation that these nuclei give rise to hyperfine splitting is just as unexpected as the hyperfine structure caused by ring protons. The finite spin density on these ^{13}C and ^{14}N nuclei is due not only to the polarization of the pair ed spins of bonding σ-electrons by the unpaired π-electron, but also to a similar interaction between the electrons in the atomic orbitals of ^{13}C and ^{14}N and this unpaired π-electron[110].

The latter interaction is thus primarily a ls-π spin polarization, though n-π spin polarization also appears in nitrogen, which has a lone pair. The complicated formulae for calculating the coupling constants $a_{C\mu}$ and $a_{N\mu}$ for ^{13}C and ^{14}N can be reduced to the simple expressions

$$a_{C\mu} = Q_C \rho_\mu^\pi + \sum_\nu Q_{X_\nu C} \rho_\nu^\pi \tag{24}$$

and

$$a_{N\mu} = Q_N \rho_\mu^\pi + \sum_\nu Q_{X_\nu N} \rho_\nu^\pi , \tag{25}$$

in which the first term on the right side is proportional to the spin population ρ_μ^π at the carbon (or the nitrogen) centre μ, while the second term depends on the spin populations ρ_ν^π at the centres ν of adjacent atoms Xν. In the radical-ions of unsubstituted aromatic hydrocarbons, the carbon centres μ are linked to two (ν and ν') or three adjacent carbon centres (ν,ν' and ν''), so that eq. (24) can be re-written as

$$a_{C\mu} = Q_C \rho_\mu^\pi + Q_{C'C}(\rho_\nu^\pi + \rho_{\nu'}^\pi) \tag{26}$$

or

$$a_{C\mu} = Q'_C \rho_\mu^\pi + Q_{C'C}(\rho_\nu^\pi + \rho_{\nu'}^\pi + \rho_{\nu''}^\pi) \tag{27}$$

The sign of the theoretically derived parameters Q_C and Q'_C is positive (+ 30 to + 40 gauss), while that of $Q_{C'C}$ is negative (ca. – 14 gauss)[110]. The sign of the coupling constant $a_{C\mu}$ can thus vary from case to case. According to eq. (26), the coupling constant a_{C1} of a ^{13}C nucleus in positions 1,4,5 and 8 of the radical-anion of naphthalene should be positive, while a_{C2} for the ^{13}C nucleus in the 2,3,6 and 7 positions should be negative (cf. Appendix A 2.1). Both predictions have been confirmed experimentally by the study of line -widths[111]. For a nitrogen contre μ linked to two carbon centres ν and ν', eq. (26) can be written analogously in the form[32,33,112-116]:

$$a_{N\mu} = Q_N \rho_\mu^\pi + Q_{CN}(\rho_\nu^\pi + \rho_{\nu'}^\pi)\,. \tag{28}$$

No quantum-mechanical calculations have been made for Q_N and Q_{CN}, and the empirical estimates put forward by different authors vary widely (cf. Section 2.2). However, it is generally accepted that $|Q_{CN}|$ is much smaller than $|Q_N|$. For this reason, the contribution to the spin populations ρ_ν^π and $\rho_{\nu'}^\pi$ is often neglected, so that eq. (28) can be written in the following form, whose simplicity is reminiscent of eq. (20) ($a_{H\mu} = Q_{CH}\ \rho_\mu^\pi$):

$$a_{N\mu} \approx Q_N \rho_\mu^\pi \tag{29}$$

The estimated magnitudes of Q_N are in the same range as those of Q_{CH}, i.e. 20-30 gauss. For equal spin polulations ρ_μ^π at the centres μ, therefore, the coupling constants $a_{N\mu}$ are of the same magnitude as coupling constants of the ring protons $a_{H\mu}$. Since theory predicts a positive value for Q_N (as it does for Q_C), the constants $a_{N\mu}$ should generally have the same sign as the spin populations ρ_μ^π at the nitrogen centres μ, i.e. the sign of $a_{N\mu}$ should also be positive. In some cases, this is supported by the study of line-widths[117]. A method for determining the sign of ^{13}C and ^{14}N coupling constants from the with of hyperfine lines will be discussed in detail in Appendix A.1.3.

Protons in alkyl substituents. The protons in the alkyl-substituted π-electron systems are denoted by α, β, γ, etc. according to whether they are linked to the π-electron systems via 0, 1, 2, etc. sp^3-hybridized carbon atoms[118]. Accordingly, the ring proton of an aromatic compound is an α-proton, the protons in a methyl and a tert-butyl substituent being respectively β- and γ-protons. In the ethyl radical the two protons on the sp^2-hybridized carbon are called α-protons, the three others in the methyl group being β-protons.

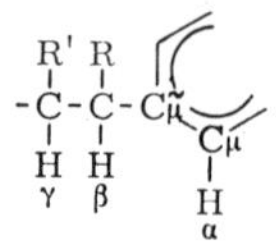

R' R
-C-C-$C_{\tilde{\mu}}$
C_{μ}
H H
γ β
H
α

The coupling constants of the α- and the β-protons of the ethyl radical, which amount to 22.4 and 26.9 gauss, respectively, are comparable in magnitude[119]. The coupling constants of the β-protons in an alkyl group at a substituted centre $\tilde{\mu}$ of an aromatic radical-ion are likewise comparable in magnitude with the coupling constants of α-protons linked directly to an unsubstituted centre μ, provided that the corresponding spin populations $\rho^{\pi}_{\tilde{\mu}}$ and ρ^{π}_{μ} are roughly equal. Beyond the β-protons, however, the coupling constants of alkyl protons rapidly decrease with increasing number of sp^3-hybridized carbon atoms situated between these protons and the π-electron system. Thus, on the assumption that $\rho^{\pi}_{\tilde{\mu}} \approx \rho^{\pi}_{\mu}$, the hyperfine splitting due to the γ-protons of an alkyl substituent at a centre μ is usually only one-thirtieth (or even less) of that caused by an α-proton at a centre μ[120]. Such small splitting frequently escapes resolution.

A great deal of experimental data has been amassed in connection with methyl-substituted radical-ions (cf. Section 2.4), and the mechanism leading to a finite spin density at β-protons has also been discussed in detail by a number of workers [51,105,121-124]. It has thus been found that the σ-π spin polarization alone is not sufficient[123] and it has been assumed that hyperconjugation is by far the most important mechanism to generate spin density at β-protons. In the MO model which takes hyperconjugation into account, the three hydrogens of the methyl substituent are regarded as a single pseudo-atom H_3. This pseudo-atom desposes over three group orbitals [125,126], which are linear combinations of the ls-functions s_I, s_{II} and s_{III} of the three hydrogens H_I, H_{II}, and H_{III}.

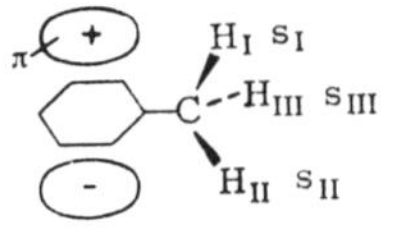

π +
H_I s_I
C--H_{III} s_{III}
-
H_{II} s_{II}

When H_{III} is in the same plane as the substituted π-electron centre, H_I and H_{II} are respectively above and below that plane. Of the three group orbitals

$$\Phi_{H_3}^{(1)} = \frac{1}{\sqrt{3}}(s_I + s_{II} + s_{III})$$

$$\Phi_{H_3}^{(2)} = \frac{1}{\sqrt{6}}(s_I + s_{II} - 2s_{III}) \tag{30}$$

and

$$\Phi_{H_3}^{(3)} = \frac{1}{\sqrt{2}}(s_I - s_{II}),$$

it is $\phi_{H_3}^{(3)}$ that has the suitable symmetry to conjugate with the π-orbitals of the system. Hyperconjugation therefore permits the unpaired electron to delocalize directly from the π-orbital into the group orbital $\phi_{H_3}^{(3)}$. Although such delocalization results only in a small spin population $\rho_{H_3}^{ls}$ in the group orbital $\phi_{H_3}^{(3)}$, this spin population $\rho_{H_3}^{ls}$ leads to coupling constants a_H of methyl protons having the correct magnitude, on account of the high density of ls-electrons at the nucleus[105, 121-123]. It is clear that $\rho_{H_3}^{ls}$ and $a_H^{CH_3}$ carry the same sign as the spin population $\rho_{\tilde{\mu}}^{\pi}$ at the substituted centre $\tilde{\mu}$, i.e., they are positive in most cases. The sign of the coupling constants of alkyl protons can be checked by NMR investigations, which have been more frequent and more successful here than in the case of ring protons[231, 275,276].

In a freely rotating methyl group, the spin polulation $\rho_{H_3}^{1s}$ is evenly distributed over the three ls-orbitals of the hydrogens, so that the relationship between $a_H^{CH_3}$ and $\rho_{H_3}^{ls}$ can be expressed as

$$a_H^{CH_3} = \frac{1}{3} Q_H \cdot \rho_{H_3}^{1s}, \tag{31}$$

where Q_H has the same size (+ 510 gauss) as the coupling constant of the proton in the hydrogen atom (see above). Eq. (31) loses its validity when the substituent ceases to rotate freely and the β-protons become more or less fixed in their positions with respect to the plane of the aromatic system. Unlike σ-π spin polarization, the extent of hyperconjugation closely depends on this position, and therefore the coupling constants of the fixed β-protons should reflect such a dependence[127]. A relationship of this kind has in fact been found for β-protons[118,128,129], which supports the assumption that hyperconjugation is the mechanism responsible for the transfer of the spin population.

A good example is the radical-anion of acenaphthene[123,130], in which the β-protons of the methylene groups are in a position that favours hyperconjugation. An

MO model analogous to (30) suggests that the coupling constant of these protons should be about 50% greater than that of the methyl protons in the radical-anion of 1,8-dimethylnaphthalene[82]. The measurements expressed below (in gauss) show that this is in fact the case.

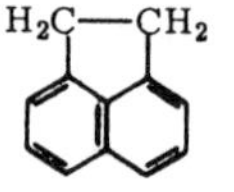

H_2C—CH_2 7,53 H_3C CH_3 4,61

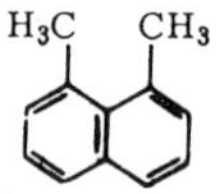

The effect of the geometry of alkyl substituents on the coupling constant of a β-proton is one of the two reasons why no relationship of general validity can be found to connect this constant and the spin population ρ_{μ}^{π} at a substituted centre $\tilde{\mu}$. The other reason is the dependence of the coupling constant on the π-charge q_{μ}^{π} at the centre $\tilde{\mu}$[80b]. This point can be best illustrated by experimental data for the radical-anion and the radical-cation derived from the same alkyl-substituted benzenoid hydrocarbon. Although these radical-ions exhibit practically the same geometry and similar spin populations ρ_{μ}^{π}, the coupling constants of the β-protons are twice as great for the cations as for the corresponding anions[51-53,131] (cf. Section 2.4, table 12).

In view of such a dependence on the π-charge q_{μ}^{π}, the approximation formula

$$a_H^{CH_3} = Q_{CCH_3}\,\rho_{\tilde{\mu}}^{\pi}, \tag{32}$$

which has often been employed with the β-protons of freely rotating methyl substituents, is only of limited use. Though the effect of the geometrical arrangement of the hydrogen atoms in the substituents is averaged out, the dependence on ρ_{μ}^{π} persists and enters into the parameter Q_{CCH_3}. Therefore, this parameter - for which theory predicts a positive sign - is not constant but behaves like a sensitive function of q_{μ}^{π}. If theory and experiment are to agree, the value of Q_{CCH_3} should vary from about +20 gauss[132] for radical-anions to about +50 gauss[80] for radical-cations.

1.6. Calculations of π-spin populations

This section deals with the MO methods applicable to the determination of the spin populations ρ_{μ}^{π}. In some of these methods such as the Hückel Molecular Orbital (HMO) and the Configuration Interaction (CI), the same models which have been used for the corresponding neutral and diamagnetic compounds are applied to radical-ions. By contrast, some other methods such as the Unrestricted Self-Consistent Field (Unrestricted SCF) and McLachlan's approximation method utilize somewhat modified models. One generally neglects the small differences in the geometrical arrangement of the π-electron centres (σ-skeleton) which can arise between neutral compounds and their radical-ions, owing to the uptake or release of a π-electron. Only exceptionally drastic changes in geometry occur when a neutral compound is converted into its radical-ion. The most famous case is that of cyclooctatetraene which will be dealt with in Section 2.5.

The HMO model. Of the various MO methods applicable to calculation of the π-spin populations, the HMO method[28,133)] is by far the simplest. It has been described in detail by several authors[28,134-136)]. The π-electrons are treated as being independent, not only of all other electrons in the system, but also of one another. The molecular orbital (MO) of a π-electron in an aromatic hydrocarbon is constructed by the linear combination (LC) of 2m atomic orbitals (AO's):

$$\Psi_i = \sum_{\mu=1}^{2m} c_{i\mu} \Phi_\mu . \tag{33}$$

MO LC AO

These AO's ϕ_μ are generally considered as $2p_z$-orbitals centred on sp^2-hybridized carbon atoms (π-electron centres). It is customary to treat the AO's ϕ_μ as normalized and orthogonal (orthonormal) functions. Orthogonality means that all overlap intergrals $<\phi_\mu|\phi_\nu>$ are zero when $\mu \neq \nu$. In an even stricter form, this is known as the Zero-Differential Overlap (ZDO) approximation (in the ZDO approximation, not only the overlap integral over the whole space, but also the corresponding differential for each volume element vanishes).

For a π-electron system consisting of 2m centres μ, the Hückel model gives 2m orthonormal MO's (HMO's) ψ_i, which are classified according to their energy levels:

$$E_i = \alpha + x_i \beta . \tag{34}$$

Since the Hückel energy parameters – the Coulomb integral α of an AO ϕ_μ and the C-C bond integral β – have a negative sign, the energy levels E_i with $x_i > 0$ are lower than those with $x_i < 0$ (the x_i values are dimensionless numbers between – 3 and +3). The HMO's ψ_i belonging to levels E_i are bonding for $x_i > 0$, and antibonding for $x_i < 0$. In general, a π-electron system consisting of 2m centres yields m bonding and m antibonding HMO's. In the HMO energy scheme for naphthalene (m = 5) shown in Figure 20, the levels of the m bonding and the m antibonding HMO's are symmetrical about the non-bonding energy E = α. For each bonding HMO ψ_i

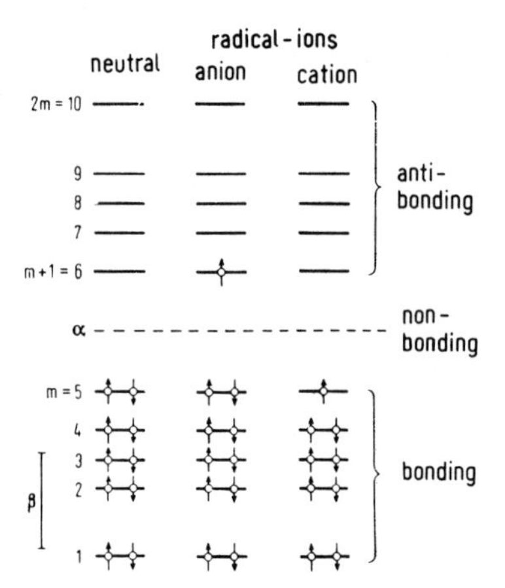

Figure 20. HMO energy scheme for an alternant system (naphthalene).

with energy $E_i = \alpha + x_i\,\beta$ there is an antibonding HMO ψ_{2m+1-i} with energy $E_{2m+1-i} = \alpha - x_i\beta$ (e.g. for ψ_1 there is ψ_{10}, for ψ_2 there is ψ_9, etc). This relationship, known as the pairing property, is exhibited by the HMO's of alternant π-electron systems, i.e. systems whose centres μ can be grouped into two sets, μ^* (starred) and μ^o (unstarred), in such a way that bonds are allowed only between centres of different sets[137]. By this definition all aromatic hydrocarbons having no odd-membered rings are alternant.

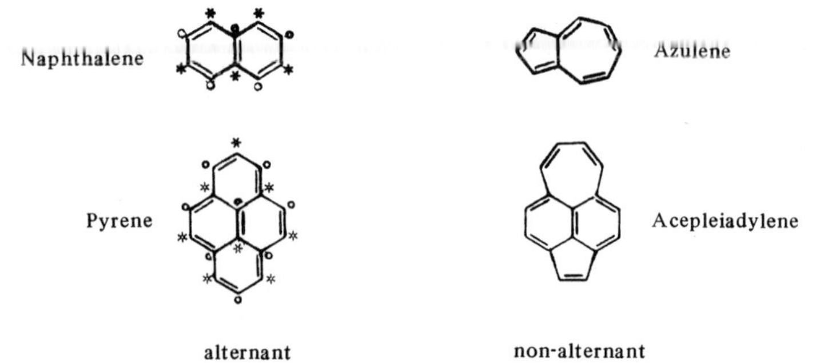

When $\psi_i = \sum_\mu c_{i\mu}\,\phi_\mu$ is an HMO of an alternant system, the HMO of $\psi_{2m+1-i} = \psi_{\hat{i}}$ can be expressed as follows:

$$\psi_{\hat{i}} = \sum_{\mu} c_{\hat{i}\mu}\,\Phi_\mu = \sum_{\mu^*} c_{i\mu^*}\,\Phi_{\mu^*} - \sum_{\mu^o} c_{i\mu^o}\,\Phi_{\mu^o} \tag{35}$$

In ψ_i and $\psi_{\hat{i}}$, the LCAO coefficients of one set are the same ($c_{\hat{i}\mu*} = c_{i\mu*}$), while those of the other set only differ in sign ($c_{\hat{i}\mu\circ} = -c_{i\mu\circ}$).

In the ground state of a neutral aromatic system, the m lowest HMO's - generally bonding - are occupied by 2m π-electrons, each HMO being occupied by two electrons with paired spins. According to eq. (18), the resultant spin density $\rho'(\mathbf{r})$ of this system is zero, because the contributions $\psi_I^2(\mathbf{r})$ coming from π-electrons with $M_S = +\frac{1}{2}$ (↑) and those from π-electrons with $M_S = -\frac{1}{2}$ (↓) cancel out. Such diamagnetic singlet states are changed into paramagnetic doublet states by addition or abstraction of a π-electron. A radical-anion is formed by introducing an extra unpaired electron into the lowest vacant HMO ψ_{m+1}, which is antibonding in most systems. On the other hand, a radical-cation is obtained when an electron is abstracted from the highest occupied (generally bonding) HMO ψ_m, leaving behind an unpaired electron in this HMO (cf. Figure 20). The two HMO's ψ_{m+1} and ψ_m will be denoted as ψ_a and ψ_b, respectively (a for antibonding and b for bonding). Since there is no interaction between π-electrons in the Hückel model, the doubly occupied HMO's do not contribute to the spin density $\rho'(\mathbf{r})$ in paramagnetic radical-ions, so that $\rho'(\mathbf{r})$ is determined solely by the contribution of the singly occupied HMO:

$$\Psi_j^2(\mathbf{r}) = \{\sum_\mu c_{j\mu}\Phi_\mu(\mathbf{r})\}^2. \tag{36}$$

where ψ_j stands for ψ_a or ψ_b, depending on whether the system in question is a radical-anion or a radical-cation.

By virtue of the normalization of ψ_j and the orthonormalization of ϕ_μ the integration of both sides of eq. (36) over space coordinates ($\mathbf{r}$) gives the simple result (cf. Appendix A 1.2):

$$1 = \sum_\mu c_{j\mu}^2. \tag{37}$$

where the squared coefficients $c_{j\mu}^2$ represent the π-spin population at the centres μ in Hückel's approximation and are called HMO spin populations. Unlike the spin populations ρ_μ^π in more refined models (in which ρ_μ^π may also be negative), the HMO values

$$c_{j\mu}^2 \approx \rho_\mu^\pi \quad (j = a \text{ or } b) \tag{38}$$

always carry a positive sign. The normalization (eq. (37)) ensures that the condition stated in eq. (21) is fulfilled.

As Figure 20 shows, the HMO's $\psi_{m+1} \equiv \psi_a$ and $\psi_m \equiv \psi_b$ of alternant systems exhibit pairing properties, and thus the following relationships apply to these systems:

$$c_{a\mu} = \pm\ c_{b\mu} \quad \text{and} \quad c^2_{a\mu} = c^2_{b\mu}\,. \tag{39}$$

The Hückel model therefore predicts equal spin populations for the radical-anion and the radical-cation of a given alternant hydrocarbon. On the other hand, the HMO's of non-alternant systems do not possess the above-mentioned pairing properties, and therefore the squared coefficients $c^2_{a\mu}$ and $c^2_{b\mu}$ of two HMO's involved are in general different. Thus on the basis of the HMO model, the radical-anion and the radical-cation of a given non-alternant hydrocarbon are expected to differ in their spin populations.

The calculated spin populations $c^2_{j\mu} \approx \rho^{\pi}_{\mu}$ at proton-carrying centres μ can be checked against experimental values with the aid of eq. (20), as is done in Section 2.1 for several radical-ions of alternant and non-alternant systems. These values confirm the predictions derived here for the radical-anion and the radical-cation of the same hydrocarbon.

Modified HMO models. The Hückel method is also used to calculate the spin populations $c^2_{j\mu}$ in aromatic radical-ions containing heteroatoms N, O, or S as π-electron centres and /or alkyl substituents. The pertubation resulting from the replacement of a carbon AO by that of an heteroatom X is expressed by two parameters h_X and k_{CX} defining the Coulomb integral α_X and the bond or resonance integral β_{CX}.

$$\alpha_X = \alpha + h_X\beta \quad \text{and} \quad \beta_{CX} = k_{CX}\beta\,. \tag{40}$$

The literature values of these parameters h_X and k_{CX} vary widely with the author and the system, but Streitwieser[28)] has proposed sets of most probable values for structurally related systems. However, these optimal sets, which are chosen to fit UV spectroscopic data and the dipole moments of neutral compounds, are not always the most suitable in the case of different experimental quantities such as ESR coupling constants of radical-ions. It has therefore been necessary to derive new sets of h_X and k_{CX} with which one can calculate HMO spin populations $c^2_{j\mu}$ that are in the best possible agreement with the observed coupling constants.

As there is a large volume of relevant experimental material some reliable estimates have been made for h_X and k_{CX} in the case of the radical-anions of pyridine-like aza-aromatic compounds (X = N) as well as semiquinone-anions (X = O). Within the following ranges, the h_X- values are somewhat larger than for the corresponding neutral compounds[28]:

$h_N = 0{,}75$ to $1{,}2$; $k_{CN} = 1{,}0$ [31–33, 58, 92, 112–116]

$h_O = 1{,}0$ to $2{,}0$; $k_{CO} = 1{,}0$ to $1{,}6$ [1, 69, 71, 138]

As with the radical-ions of hydrocarbons, eq. (20) again permits the HMO spin populations $c_{j\mu}^2 \approx \rho_\mu^\pi$ at proton-carrying carbon centres μ to be compared with the experimental values $a_{H\mu}$. For checking the calculated spin populations at sp^2-hybridized nitrogens, one can use eq. (28) or the approximation formula (29), which connects these spin populations with the coupling constants $a_{N\mu}$ of ^{14}N nuclei. By contrast, it is more difficult to check the spin populations at oxygen and sulphur centres in a direct manner, since the most common isotopes ^{16}O and ^{32}S are not magnetic. Only a few recent reports[283] deal with the hyperfine structure of radical-anions enriched in the magnetic ^{17}O nucleus. The splittings arising from the magnetic ^{33}S isotope in natural abundance have also been recently investigated for a few radical-cations[316].

The spin populations in alkyl-substituted radical-anions are easiest to calculate with the aid of an HMO method that takes into account only the inductive effect of the substituents. To do this, it is generally sufficient to modifiy the Coulomb integral $\alpha_{\tilde{\mu}}$ of the substituted π-electron centre $\tilde{\mu}$[28]:

$$\alpha_{\tilde{\mu}} = \alpha + h_{\tilde{\mu}}\beta . \tag{41}$$

In the case of the radical-anions, $h_{\tilde{\mu}}$ values of −0.3 seem to reproduce correctly the changes in the HMO spin populations brought about by the inductive effect of the alkyl substituents[82,116,132]. Eq. (20) can be used here to check the agreement between the experimental values $a_{H\mu}$ and the spin populations $c_{j\mu}^2 \approx \rho_\mu^\pi$ at unsubstituted centres μ carrying ring protons (α-protons). The approximation formula (32) correlates the coupling constants $a_H^{CH_3}$ of the β-protons in freely rotating methyl groups with the spin populations $c_{j\tilde{\mu}}^2 \approx \rho_{\tilde{\mu}}$ at the substituted centres $\tilde{\mu}$. As has been stressed in Section 1.5, however, the dependence of the parameter Q_{CCH_3} on the π-charge $q_{\tilde{\mu}}^\pi$ must not be neglected in this connection.

The method involving only the inductive effect, of course, does not yield any spin population in the alkyl substituents themselves. To obtain such a population in a freely rotating methyl substituent, one can use a hyperconjugative HMO model, in which this substituent is treated as a two-centre grouping:

$$\rangle C_{\tilde{\mu}} - C_M \equiv H_3 .$$

The particular nature of such a grouping is taken into account by the choice of suitable Coulomb and bond integrals[1,82,105,126)]

$$\begin{aligned} \alpha_M &= \alpha + h_M\beta \quad \text{with} \quad h_M = 0 \ \text{or} \ -0{,}1; \\ \alpha_{H_3} &= \alpha + h_{H_3}\beta \quad \text{with} \quad h_{H_3} = 0 \ \text{to} \ -0{,}5; \end{aligned} \tag{42}$$

$$\begin{aligned} \beta_{\bar{\mu}M} &= k_{\bar{\mu}M}\beta \quad \text{with} \quad k_{\bar{\mu}M} = 0{,}7 \ \text{to} \ 1{.}0; \\ \beta_{MH_3} &= k_{MH_3}\beta \quad \text{with} \quad k_{MH_3} = 2{,}5 \ \text{or} \ 3{,}0\,. \end{aligned} \tag{43}$$

The HMO spin population $c^2_{jH_3}$ in the group orbital $\phi^{(3)}_{H_3}$, given by such a model, is a good approximation for $\rho^{ls}_{H_3}$. (This spin population was denoted in Section 1.5 as $\rho^{ls}_{H_3}$ and not as $\rho^{\pi}_{H_3}$, because the unpaired electron delocalizes into the group orbital $\phi^{(3)}_{H_3}$, which is a linear combination of the ls-orbitals of the hydrogens.) Eq. (31) correlates $c^2_{jH_3}$ ($\approx \rho^{ls}_{H_3}$) with the coupling constants $a^{CH_3}_H$ of the methyl protons.

Refined methods. The HMO models often give $c^2_{j\mu}$ values that are zero or very small, although the experimental data suggest appreciable spin populations ρ^{π}_{μ} at the centres μ in question. This discrepancy reveals the limitations of a one-electron method. The HMO models fail by not considering the π-π spin polarization. As mentioned in Section 1.5, this polarization is an interaction between the unpaired electron spin and the paired spins of the other π-electrons. It leads to negative spin populations ρ^{π}_{μ} at the centres μ, for which the HMO approximation values $c^2_{j\mu}$ are vanishingly small.

The π-π spin polarization can be illustrated with the aid of a simple scheme involving three π-electrons and two adjacent centres of an aromatic radical-ion, this scheme being analogous to that used in connection with the σ-π spin polarization in a C-H section (cf. Figure 19 and Section 1.5). As usual, the unpaired π-electron with a spin quantum number $M_S = +\frac{1}{2}$ will occupy the HMO $\psi_j = \sum_{\mu} c_{j\mu}\,\phi_{\mu}$. It is assumed that this electron spends most of its time at the centre $\mu = 1$ (c^2_{j1} is large), while its residence probability at centre $\mu = 2$ is vanishingly small ($c^2_{j2} \approx 0$). Therefore, the unpaired electron is assumed to be at centre $\mu = 1$ in the scheme. The other two π-electrons in question are paired, and occupy an HMO ψ_i. When they are shared by the centres 1 and 2, the following spin orientations can arise:

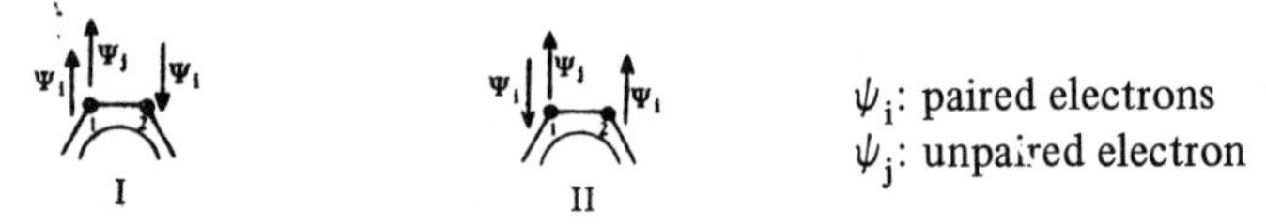

In arrangement I, the paired electron which is at centre 1 has the same spin quantum number as the unpaired electron ($M_S = +\frac{1}{2}, \uparrow$), while its partner at centre

2 will have the opposite spin ($M_S = -\frac{1}{2}, \downarrow$). In arrangement II, on the other hand, the spins of the two paired electrons at both centres are reversed, so that the one at centre 1 then differs in its spin quantum number from the unpaired electron. According to Hund's rule (cf. Section 1.5) arrangement I is preferred. This means that the paired electron with $M_S = +\frac{1}{2}$ will be more often at centre 1 than its partner with $M_S = -\frac{1}{2}$. It is clear that the opposite must be true for centre 2. The generalized conclusion is as follows: paired π-electrons having the same spin quantum number ($M_S = +\frac{1}{2}$) as the unpaired π-electron favour centres μ with high $c_{j\mu}^2$ values, while their partners having the opposite spin quantum number ($M_S = -\frac{1}{2}$) are shifted to centres μ with low $c_{j\mu}^2$ values. As a consequence, the π-spin populations will be enhanced at centres with a high $c_{j\mu}^2$ value by positive contributions, whereas the π-spin populations will be reduced at centres with a low $c_{j\mu}^2$ value by negative contributions. If the magnitude of the latter exceeds that of $c_{j\mu}^2$, then negative spin populations ρ_μ^π should result. Refined MO methods, that take into account the π-π spin polarization, indeed indicate negative values of ρ_μ^π for these centres μ. Such methods include the Configuration Interaction (CI) and the Unrestricted Self-Consistent Field (Unrestricted SCF) method. It will be shown in Appendix A 1.2 how the spin populations ρ_μ^π can be calculated with the aid of the former. As regard the Unrestricted SCF method, it suffices to say here that it represents a modification of the ordinary or restricted SCF method, used mainly to calculate the orbital energies first of atoms (Hartree-Fock)[140], and then of π-electron systems[26,141,142]. The unrestricted version is preferred for the calculation of ρ_μ^π, because it takes into account the π-π spin polarization. This is achieved by neglecting the restriction that any two paired π-electrons in the radical-ion must occupy one and the same orbital (closed shell)[143]. When these electrons are assumed to have somewhat differing orbitals, their contributions to the spin density $\rho'(\boldsymbol{r})$ no longer cancel out, and thus negative π-spin populations may arise[144] (i.e. the two electrons are no longer perfectly paired). According to a method developed by McLachlan[145], the π-spin populations can be calculated as

$$\rho_\mu^\pi = c_{j\mu}^2 + \lambda \sum_\nu \pi_{\mu\nu} c_{j\nu}^2 \tag{44}$$

where $c_{j\mu}^2$ and $c_{j\nu}^2$ are the HMO values for the π-spin populations at the centre μ and at the other centres ν in the system, respectively. Furthermore, the $\pi_{\mu\nu}$ terms represent the HMO polarizabilities of the neutral diamagnetic system, being defined as[28,137]:

$$\pi_{\mu\nu} = 4 \sum_{i=1}^{m} \sum_{l=m+1}^{2m} \frac{c_{i\mu} c_{l\mu} c_{i\nu} c_{l\nu}}{x_i - x_l}, \tag{45}$$

where the subscript i (= 1, . . . , m) refers to the doubly occupied and generally bonding HMO's while the subscript ℓ (= m + 1. . . 2m) refers to the m unoccupied and generally antibonding HMO's. The value of the parameter λ in McLachlan's equation (eq. (44)) has been estimated[145] at 1.2, which gives good agreement in the case of hydrocarbon radical-ions. Although McLachlan's formula is based on the Unrestriced SCF method, a comparison of eqs. (44) and (45) with the corresponding expressions in Appendix A 1.2 shows that it is analogous to the CI method.

Despite its relative simplicity, eq. (44) gives spin populations ρ_μ^π comparable to those obtained by the more exact Unrestricted SCF or the CI method (cf. Appendix A 1.2). Moreover, this equation gives a clear expression of the π-spin polarization which leads to negative ρ_μ^π values.

It should be noted finally that the two complementary HMO statements about the radical-anion and the radical-cation of a given hydrocarbon remain valid in refined MO models that explicitly include the π-electronic interactions but retain the ZDO approximation[148]. Thus, in this approximation, which is the one most often used, both the CI and the unrestricted SCF methods postulate that the radical-anion and the radical-cation derived from a given system have equal spin populations ρ_μ^π when the compound is alternant, and different spin populations when the compound is non-alternant.

Appendix to Part 1

A.1.1. σ-π Spin polarization

It will now be shown how the configuration interaction (CI) method of the MO theory can be used to determine the σ-π spin polarization in the C-H section of an aromatic radical-ion[104].

If one confines the consideration to the unpaired π-electron and the two σ-electrons of a C-H bond, these σ-electrons being paired in the ground configuration (cf. Figure 19), then one need consider here only the contribution of those singly excited configurations which arise by the promotion of an electron from a bonding σ-orbital to an antibonding σ^*-orbital. It can be easily shown that all other singly excited configurations (e.g. those resulting from promotions $\sigma \rightarrow \pi$, $\pi \rightarrow \sigma^*$ and $\pi \rightarrow \pi^*$) do not contribute to the σ-π spin polarization, either because they do not mix with the ground configuration for symmetry reasons or because the σ-electrons in these configurations remain paired.

The ground configuration and the singly excited configurations resulting from the $\sigma \longrightarrow \sigma^*$ promotion are represented as Slater determinants, which contain not only the orbitals but also the spin functions (α or β) of the three electrons. In the following, the presence and the absence of a bar over the symbol for the orbital will denote the β spin function ($M_S = -\frac{1}{2}$) and the α spin function ($M_S = +\frac{1}{2}$), respectively.

Using the Slater determinants, one constructs the quartet configuration ($S^{tot} = 3/2$; $2S^{tot} + 1 = 4$) and the doublet configurations ($S^{tot} = \frac{1}{2}$; $2S^{tot} + 1 = 2$) which are represented here by their components $M_S^{tot} = +\frac{1}{2}$. The determinant

$$\Delta_0 = |\sigma\bar{\sigma}\pi| = \frac{1}{\sqrt{3!}} \begin{vmatrix} \sigma^{(1)}\bar{\sigma}^{(1)}\pi^{(1)} \\ \sigma^{(2)}\bar{\sigma}^{(2)}\pi^{(2)} \\ \sigma^{(3)}\bar{\sigma}^{(3)}\pi^{(3)} \end{vmatrix} \tag{46}$$

in which the superscripts refer to the coordinates of the three electrons, stands for the doublet ground configuration $^2\chi_0$. The three determinants

$$\begin{aligned} \Delta_1 &= |\sigma\bar{\sigma}^*\pi| \\ \Delta_2 &= |\bar{\sigma}\sigma^*\pi| \\ \Delta_3 &= |\sigma\sigma^*\bar{\pi}| \end{aligned} \tag{47}$$

are used analogously to denote the excited quartet configuration

$$^{4}\chi_1 = \frac{1}{\sqrt{3}}(\Delta_1 + \Delta_2 + \Delta_3) = \frac{1}{\sqrt{3}}(|\sigma\bar{\sigma}^*\pi| + |\bar{\sigma}\sigma^*\pi| + |\sigma\sigma^*\bar{\pi}|) \tag{48}$$

and the two excited doublet configurations

$$^{2}\chi_1 = \frac{1}{\sqrt{6}}(\Delta_1 + \Delta_2 - 2\Delta_3) = \frac{1}{\sqrt{6}}(|\sigma\bar{\sigma}^*\pi| + |\bar{\sigma}\sigma^*\pi| - 2|\sigma\sigma^*\bar{\pi}|) \tag{49}$$

$$^{2}\chi_1' = \frac{1}{\sqrt{2}}(\Delta_1 - \Delta_2) = \frac{1}{\sqrt{2}}(|\sigma\bar{\sigma}^*\pi| - |\bar{\sigma}\sigma^*\pi|) \tag{50}$$

arising from the $\sigma \longrightarrow \sigma^*$ promotion (cf. Figure 21).

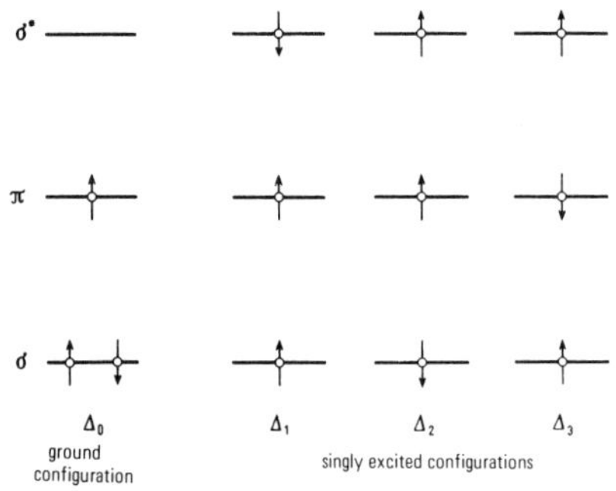

σ*
π
σ
Δ₀ ground configuration
Δ₁ Δ₂ Δ₃ singly excited configurations

Figure 21: Ground configuration and excited ($\sigma \longrightarrow \sigma^*$) configurations for a C-H section of an aromatic radical ion (see text).

Since only configurations of the same multiplicity mix, in first-order perturbation theory, the ground state function is given by

$$^{2}\Gamma_0 = {}^{2}\chi_0 + \lambda\,{}^{2}\chi_1 + \lambda'\,{}^{2}\chi_1', \tag{51}$$

where the coefficients $|\lambda|$, $|\lambda'| \ll 1$ give a measure of the mixing of $^2\chi_0$ with $^2\chi_1$ and $^2\chi_1'$, respectively.

At this point the spin density operator $\mathfrak{S}^{tot}$

$$\mathfrak{S}^{tot} = \sum_k \mathfrak{S}^{(k)}$$

should be introduced.

It is constructed from the one-electron operators $\mathfrak{S}^{(k)}$ acting exclusively on the spin part of the spin-orbital function of the k-th electron. In the case of a spin-orbital function $\phi_j(\mathbf{r})\,\alpha(\omega)$ or $\phi_j(\mathbf{r})\,\beta(\omega)$, where $\mathbf{r}$ and ω are the space and the spin coordinates respectively, such an operation may be formulated as follows:

$$\begin{aligned} \mathfrak{S}\,\Phi_j(\mathrm{r})\,\alpha(\omega) &= +\,\Phi_j(\mathrm{r})\,\alpha(\omega) \\ \mathfrak{S}\,\Phi_j(\mathrm{r})\,\beta(\omega) &= -\,\Phi_j(\mathrm{r})\,\beta(\omega)\,, \end{aligned} \tag{52}$$

i.e. the task of the operator σ is simply to give a positive or a negative sign to the orbital $\phi_j(\mathbf{r})$, according to the spin quantum number M_S. (The operator $\mathfrak{S}$ is usually represented as the twofold product of the z component of the spin operator, $2\,S_z$, and the Dirac δ function which yields the value of ϕ_j at the position $\mathbf{r}$.)

The multiplication of (52) from the left by a second spin-orbital function $\phi_i(\mathbf{r})\,\alpha(\omega)$ or $\phi_i(\mathbf{r})\,\beta(\omega)$ and integration over the spin coordinates ω result in*)

$$\begin{aligned} \int \Phi_i(\mathrm{r})\,\alpha(\omega)\,\mathfrak{S}\,\Phi_j(\mathrm{r})\,\alpha(\omega)\,\mathrm{d}\omega &= +\,\Phi_i(\mathrm{r})\,\Phi_j(\mathrm{r}) \\ \int \Phi_i(\mathrm{r})\,\alpha(\omega)\,\mathfrak{S}\,\Phi_j(\mathrm{r})\,\beta(\omega)\,\mathrm{d}\omega &= 0 \\ \int \Phi_i(\mathrm{r})\,\beta(\omega)\,\mathfrak{S}\,\Phi_j(\mathrm{r})\,\alpha(\omega)\,\mathrm{d}\omega &= 0 \\ \int \Phi_i(\mathrm{r})\,\beta(\omega)\,\mathfrak{S}\,\Phi_j(\mathrm{r})\,\beta(\omega)\,\mathrm{d}\omega &= -\,\Phi_i(\mathrm{r})\,\Phi_j(\mathrm{r}). \end{aligned} \tag{53}$$

These integrals can be abbreviated as

$$\begin{aligned} [\Phi_i\,|\,\mathfrak{S}\,|\,\Phi_j] &= +\,\Phi_i\,\Phi_j \\ [\Phi_i\,|\,\mathfrak{S}\,|\,\bar{\Phi}_j] &= 0 \\ [\bar{\Phi}_i\,|\,\mathfrak{S}\,|\,\Phi_j] &= 0 \\ [\bar{\Phi}_i\,|\,\mathfrak{S}\,|\,\bar{\Phi}_j] &= -\,\Phi_i\,\Phi_j. \end{aligned} \tag{53a}$$

The resulting product ($+\phi_i\phi_j$ or $-\phi_i\phi_j$), which is still a function of the space vector r, represents the contribution of the k-th electron to the spin density $\rho'(\mathbf{r})$. The total spin density of a system comprising k electrons is obtained by applying the operator $\mathfrak{S}^{\text{tot}} = \sum_k \mathfrak{S}^{(k)}$. The integration now yields a sum of integrals resulting from the consecutive action of each operator $\mathfrak{S}^{(k)}$ on the multi-electron function in turn. The square brackets indicate that for the electron k, on which an operator $\mathfrak{S}^{(k)}$ is acting, the integration is carried out only over the spin coordinates, whereas for all the other electrons it is over both the spin and the space coordinates. This method gives the following result for the spin density $\rho'(\mathbf{r})$ in the ground configuration $^2\chi_0 = \Delta_0$ [cf. eq. (46)].

$$[{}^2\chi_0\,|\,\mathfrak{S}^{\text{tot}}\,|\,{}^2\chi_0] = [\sigma\,|\,\mathfrak{S}\,|\,\sigma] + [\bar{\sigma}\,|\,\mathfrak{S}\,|\,\bar{\sigma}] + [\pi\,|\,\mathfrak{S}\,|\,\pi] = \sigma^2 - \sigma^2 + \pi^2 = \pi^2. \tag{54}$$

---

*) It is assumed here that all the orbital functions are real. If they are not, then one orbital function in a product of two must be replaced by its complex conjugate.

Owing to the opposite signs of the spin quantum numbers M_S, the first two terms on the right cancel out. The result fully agrees with the expectation that spin density in the ${}^2\chi_0$ configuration should have a pure π-character.

The use of the ground state function ${}^2\Gamma_0$(eq. (51)) leads in a first-order approximation to

$$[{}^2\Gamma_0|\mathfrak{S}^{tot}|{}^2\Gamma_0] = [{}^2\chi_0|\mathfrak{S}^{tot}|{}^2\chi_0] + 2\lambda[{}^2\chi_0|\mathfrak{S}^{tot}|{}^2\chi_1] + 2\lambda'[{}^2\chi_0|\mathfrak{S}^{tot}|{}^2\chi_1'], \tag{55}$$

where the first term on the right is identical with eq. (54). The last two terms yield the integrals (cf. eqs. (49) and (50))

$$[{}^2\chi_0|\mathfrak{S}^{tot}|{}^2\chi_1] = \frac{1}{\sqrt{6}}\{[\Delta_0|\mathfrak{S}^{tot}|\Delta_1] + [\Delta_0|\mathfrak{S}^{tot}|\Delta_2] - 2[\Delta_0|\mathfrak{S}^{tot}|\Delta_3]\} \tag{56}$$

and

$$[{}^2\chi_0|\mathfrak{S}^{tot}|{}^2\chi_1'] = \frac{1}{\sqrt{2}}\{[\Delta_0|\mathfrak{S}^{tot}|\Delta_1] - [\Delta_0|\mathfrak{S}^{tot}|\Delta_2]\}. \tag{57}$$

It can easily be shown that by using expressions (46) and (47) for Δ_0, Δ_1, Δ_2, and Δ_3, the following equations result:

$$[\Delta_0|\mathfrak{S}^{tot}|\Delta_1] = [\bar{\sigma}|\mathfrak{S}|\bar{\sigma}^*] = -\sigma\sigma^* \tag{58}$$

$$[\Delta_0|\mathfrak{S}^{tot}|\Delta_2] = -[\sigma|\mathfrak{S}|\sigma^*] = -\sigma\sigma^* \tag{59}$$

$$[\Delta_0|\mathfrak{S}^{tot}|\Delta_3] = 0. \tag{60}$$

The addition or subtraction of these integrals in accordance with eqs. (56) and (57) then leads respectively to

$$[{}^2\chi_0|\mathfrak{S}^{tot}|{}^2\chi_1] = -\frac{2}{\sqrt{6}}\sigma\sigma^* \tag{61}$$

$$[{}^2\chi_0|\mathfrak{S}^{tot}|{}^2\chi_1'] = 0. \tag{62}$$

Substitution of [eq. (61) and (62)] into (55) gives the spin density in ${}^2\Gamma_0$:

$$[{}^2\Gamma_0|\mathfrak{S}^{tot}|{}^2\Gamma_0] = \pi^2 - \frac{4}{\sqrt{6}}\lambda\sigma\sigma^*. \tag{63}$$

This spin density

$$\rho'(\mathbf{r}) = \pi^2(\mathbf{r}) - \frac{4}{\sqrt{6}}\lambda\sigma(\mathbf{r})\sigma^*(\mathbf{r})$$

differs from the spin density in $^2\chi_0$

$$\rho'(\mathbf{r}) = \pi^2(\mathbf{r}) \tag{54}$$

by the small σ-term ($|\lambda| \ll 1$)

$$-\frac{4}{\sqrt{6}}\lambda\,\sigma(\mathbf{r})\,\sigma^*(\mathbf{r}), \tag{64}$$

representing the contribution of the excited configuration $^2\chi_1$. By contrast, the second excited configuration $^2\chi_1'$ does not contribute anything to the spin density in $^2\Gamma_0$.

The π-term vanishes at the site of the proton ($\mathbf{r} = 0$), so that all the spin density $\rho'(0)$ at this location is due to the σ -term (64).

$$\rho'(0) = -\frac{4}{\sqrt{6}}\lambda\,\sigma(0)\,\sigma^*(0). \tag{65}$$

Insertion of eq. (65) into

$$a_H = K_H\rho'(0) \quad (K_H = 2{,}3626 \times 10^{-22}\ \text{gauss cm}^3) \tag{14}$$

yields

$$a_{H\mu} = -\frac{4}{\sqrt{6}}K_H\lambda\,\sigma(0)\,\sigma^*(0), \tag{66}$$

where $a_{H\mu}$ is the coupling constant of the ring proton Hμ.

The coefficient λ of the linear combination (eq. (51)) to which the coupling constant $a_{H\mu}$ is proportional, depends in the first order approximation on the cross term between the configurations $^2\chi_1$ and $^2\chi_0$, as well as on the difference between the energies of these (E_1 and E_0).

$$\lambda = -\frac{\int {}^2\chi_0\,\mathscr{H}^{tot}\,{}^2\chi_1\,d\tau}{E_1 - E_0} \equiv -\frac{\langle {}^2\chi_0|\mathscr{H}^{tot}|{}^2\chi_1\rangle}{E_1 - E_0} \tag{67}$$

The Hamiltonian operator $\mathscr{H}^{tot}$ is split as usual into one-electron and two-electron operators:

$$\mathscr{H}^{tot} = \sum_k \mathscr{H}^{(k)} + \sum_{k<l} \mathfrak{G}^{(k,l)},$$

where $\mathscr{H}^{(k)}$ denotes the kinetic and the potential energy of the k-th electron, while $\mathfrak{G}^{(kl)}$ ($= e^2/r_{kl}$) is the electrostatic repulsion between the k-th and the l-th electron. Furthermore, τ comprises both space and the spin coordinates. The integral $\langle {}^2\chi_0 | \mathscr{H}^{tot} | {}^2\chi_1 \rangle$ gives, in accordance with eqs. (46) and (49), the following expression:

$$\langle {}^2\chi_0 | \mathscr{H}^{\text{tot}} | {}^2\chi_1 \rangle = \frac{1}{\sqrt{6}} \{ \langle \Delta_0 | \mathscr{H}^{\text{tot}} | \Delta_1 \rangle + \langle \Delta_0 | \mathscr{H}^{\text{tot}} | \Delta_2 \rangle - 2 \langle \Delta_0 | \mathscr{H}^{\text{tot}} | \Delta_3 \rangle \} \tag{68}$$

along with

$$\langle \Delta_0 | \mathscr{H}^{\text{tot}} | \Delta_1 \rangle = \langle \bar{\sigma} | \mathscr{H} | \bar{\sigma}^* \rangle + \langle \sigma \bar{\sigma} | \mathfrak{G} | \sigma \bar{\sigma}^* \rangle + \langle \bar{\sigma} \pi | \mathfrak{G} | \bar{\sigma}^* \pi \rangle \tag{69}$$

$$\langle \Delta_0 | \mathscr{H}^{\text{tot}} | \Delta_2 \rangle = -\langle \sigma | \mathscr{H} | \sigma^* \rangle - \langle \sigma \bar{\sigma} | \mathfrak{G} | \sigma^* \bar{\sigma} \rangle - \langle \sigma \pi | \mathfrak{G} | \sigma^* \pi \rangle + \langle \sigma \pi | \mathfrak{G} | \pi \sigma^* \rangle \tag{70}$$

and $$\langle \Delta_0 | \mathscr{H}^{\text{tot}} | \Delta_3 \rangle = -\langle \bar{\sigma} \pi | \mathfrak{G} | \bar{\pi} \sigma^* \rangle . \tag{71}$$

In the one-electron and the two-electron integrals of eqs. (69), (70), and (71), the integrals containing the same orbital parts are equal, e.g.

$$\langle \sigma | \mathscr{H} | \sigma^* \rangle = \langle \bar{\sigma} | \mathscr{H} | \bar{\sigma}^* \rangle$$

and $$\langle \sigma \pi | \mathfrak{G} | \pi \sigma^* \rangle = \langle \bar{\sigma} \pi | \mathfrak{G} | \bar{\pi} \sigma^* \rangle .$$

Summation of (69), (70), and (71) thus yields the following integral for the cross term between ${}^2\chi_0$ and ${}^2\chi_1$ (eq. (68)):

$$\begin{aligned} \langle {}^2\chi_0 | \mathscr{H}^{\text{tot}} | {}^2\chi_1 \rangle &= +\frac{3}{\sqrt{6}} \langle \sigma \pi | \mathfrak{G} | \pi \sigma^* \rangle \\ &\equiv +\frac{3}{\sqrt{6}} \int \sigma^{(1)} \pi^{(2)} \frac{e^2}{r_{12}} \pi^{(1)} \sigma^{*(2)} \, d\tau . \end{aligned} \tag{72}$$

The difference between the energies of ${}^2\chi_1$ and ${}^2\chi_0$, i.e. $E_1 - E_0$, can be roughly equated to $E_\sigma^* - E_\sigma$, i.e. to the promotion energy of an electron from the bonding σ-orbital to an antibonding σ^*-orbital. The expression for λ is thus of the form

$$\lambda = -\frac{3}{\sqrt{6}} \frac{\langle \sigma \pi | \mathfrak{G} | \pi \sigma^* \rangle}{E_{\sigma^*} - E_\sigma} \tag{73}$$

When this formula is used for λ in eq. (66), the final result is

$$a_{H\mu} = +2 K_H \frac{\langle \sigma \pi | \mathfrak{G} | \pi \sigma^* \rangle}{E_{\sigma^*} - E_\sigma} \sigma(0) \sigma^*(0) \tag{74}$$

The molecular orbital π of the unpaired electron is now approximated as

$$\pi \approx \psi_j = \sum_\nu c_{j\nu} \Phi_\nu , \tag{75}$$

which is a linear combination of $2p_z$ atomic orbitals ϕ_ν of those carbon centres ν over which the π-electron is delocalized (LCAO-MO: cf. Section 1.6).

In the ZDO approximation, the integrals $\langle \sigma \phi_\nu \mid \mathfrak{G} | \phi_{\nu'} \sigma^* \rangle$ with $\nu \neq \nu'$ can be neglected, so that eq. (74) has the form

$$a_{H\mu} = 2\,K_H \frac{\sum_{\nu} c_{j\nu}^2 \langle \sigma \Phi_\nu | \mathfrak{G} | \Phi_\nu \sigma^* \rangle}{E_{\sigma^*} - E_\sigma} \sigma(0)\,\sigma^*(0) \tag{76}$$

It can be further expected that the contribution of only that carbon atom which participates in the $C\mu$-$H\mu$ bond ($\nu = \mu$) is important in eq. (76), the contribution of all other centres ($\nu \neq \mu$) being negligibly small.

This fact enables one to replace the summation over ν by a single term ($\nu = \mu$). After a slight rearrangement, the formula for the coupling constant $a_{H\mu}$ of a ring proton is then

$$a_{H\mu} = \underbrace{\left\{ 2\,K_H \frac{\langle \sigma \Phi_\mu | \mathfrak{G} | \Phi_\mu \sigma^* \rangle}{E_{\sigma^*} - E_\sigma} \sigma(0)\,\sigma^*(0) \right\}}_{Q_{CH}} c_{j\mu}^2 \,. \tag{77}$$

The expression covered by the horizontal brace in eq. (77) is fundamentally constant for aromatic radical-ions, so that there is a linear relationship between $a_{H\mu}$ and the theoretical quantity $c_{j\mu}^2$. Such a relationship was first proposed by McConnell[108] on an empirical basis (cf. Section 1.5). The parameter Q_{CH} in McConnell's equation (20) is identical with the braced expression picked out in eq. (77). The numerical values estimated for this empirically determined expression are in good agreement with the value of $|Q_{CH}|$(20-30 gauss). This agreement confirms the validity of the proposed σ-π polarization mechanism[107]. An analogous mechanism, the π-π spin polarization[107,109] – which is not considered in Appendix A1.1 – results in the necessity of replacing the squares $c_{j\mu}^2$ of the LCAO coefficients in eq. (77) by the π-spin population ρ_μ^π (cf. Appendix A 1.2). Eq. (77) is thus transformed into eq. (20);

$$a_{H\mu} = Q_{CH}\,\rho_\mu^\pi \tag{20}$$

The numerical values of Q_{CH} obtained for the braced expression in eq. (77) will not be given here. Instead, it will be shown as simply as possible that Q_{CH} in eq. (77) has a negative sign. For this purpose, it is sufficient to express the molecular orbitals σ and σ^* as a crude approximation by

$$\begin{aligned} \sigma &= (1/\sqrt{2})(t + 1s) \\ \sigma^* &= (1\sqrt{2})(t - 1s)\,, \end{aligned} \tag{78}$$

where t and ls represent the atomic orbitals respectively of the carbon and the hydrogen flanking the $C\mu$-$H\mu$ bond in question; t denotes the trigonal sp^2-hybrid orbital of the carbon, and ls the ground-state hydrogen wave function. One thus obtains:

$$Q_{CH} \approx \frac{K_H \{\langle t\,\Phi_\mu | \mathfrak{H} | \Phi_\mu t\rangle - \langle 1s\,\Phi_\mu | \mathfrak{H} | \Phi_\mu 1s\rangle\}}{2(E_{\sigma^*} - E_\sigma)} \tfrac{1}{2}\{t^2(0) - 1s^2(0)\}. \tag{79}$$

The following relationships a) - c) hold in this connection:

a) $\langle t\,\Phi_\mu | \mathfrak{H} | \Phi_\mu t\rangle > \langle 1s\,\Phi_\mu | \mathfrak{H} | \Phi_\mu 1s\rangle$.

The first integral is larger, since AO's t and ϕ_μ in it are centred on the same atom $C\mu$ while the AO's ϕ_μ and ls in the second integral belong to different atoms (i.e. to $C\mu$ and $H\mu$).

b) $E_{\sigma^*} > E_\sigma$.

The energy of the antibonding MO σ^* is, of course, higher than that of the bonding MO σ. And finally

c) $t^2(0) \ll 1s^2(0)$,

i.e. the square of the carbon function t at the site of the proton (0) is vanishingly small in comparison with the square of the hydrogen function 1s at the same site. In eq. (79), therefore, the differences between the two integrals in the numerator and between the two energies in the denominator are positive, but the difference between the electron densities at the proton is negative. Since K_H (= 2.3626 x 10^{-22} gauss · cm^3), is a positive constant, the parameter Q_{CH} has a negative sign.

A.1.2. π-π Spin polarization

The π-π spin polarization can be taken into account by the configuration interaction (CI) method in a manner analogous to that used to allow for the σ-π spin polarization[147]. The ground configuration $^2\chi_0$ and the singly excited configurations $^2\chi_1$, whose interaction with $^2\chi_0$ determines the extent of the π-π spin polarization, are again represented as Slater determinants, which this time, however, comprise only π-functions (generally HMO's). The ground configuration with $M_S^{tot} = +\frac{1}{2}$ is represented as

$$^2\chi_0 = |\psi_1 \bar{\psi}_1 \cdots \psi_i \bar{\psi}_i \cdots\cdots \psi_{j-1} \bar{\psi}_{j-1} \psi_j|, \tag{80}$$

where each HMO ψ_i (i = 1, . . . , j-1) is occupied by two electrons differing in their spin quantum number M_S, while the unpaired electron is in the HMO ψ_j (cf. Figure 22).

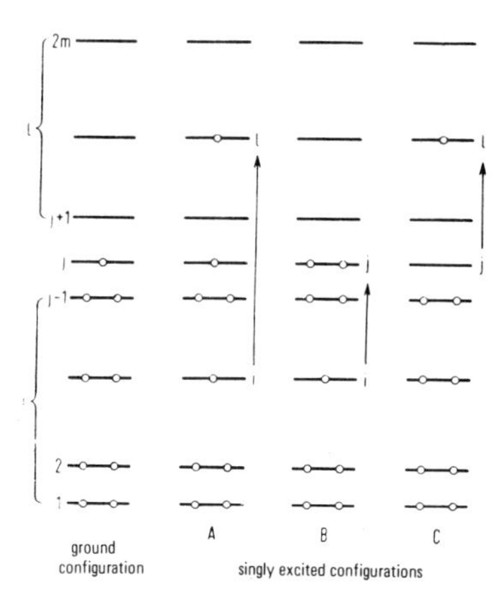

Figure 22: Ground configuration and the singly excited ($\pi \rightarrow \pi^*$) configurations of a radical-ion.

Orbital subscript:	i = 1 to j-1	j	l = j+1 to 2m
Occupancy in ground configuration:	two	one	none

Figure 22 illustrates the following modes of promoting a π-electron from the ground configuration:

A. $i \longrightarrow l$ (from the doubly occupied HMO ψ_i to the vacant one ψ_l)
B. $i \longrightarrow j$ (from the doubly occupied HMO ψ_i to the singly occupied one ψ_j)
C. $j \longrightarrow l$ (from the singly occupied HMO ψ_j to the vacant one ψ_l)

As in the case of σ-π interaction, each of the $i \longrightarrow l$ transitions leads to a quartet configuration $^4\chi_{il}$ and to two doublet configurations $^2\chi_{il}$ and $^2\chi_{il}$

$$^2\chi_{il} = \frac{1}{\sqrt{6}}\{|\ldots\ldots\psi_i\bar{\psi}_l\ldots\ldots\psi_j| + |\ldots\ldots\bar{\psi}_i\psi_l\ldots\ldots\psi_j| - 2|\ldots\ldots\psi_i\psi_l\ldots\ldots\bar{\psi}_j|\} \tag{81}$$

$$^2\chi'_{il} = \frac{1}{\sqrt{2}}\{|\ldots\ldots\psi_i\bar{\psi}_l\ldots\ldots\psi_j| - |\ldots\ldots\bar{\psi}_i\psi_l\ldots\ldots\psi_j|\} \tag{82}$$

The other two types of transitions yield only doublet configurations

$$^2\chi_{ij} = |\ldots\psi_i\bar{\psi}_j\ldots\psi_j| \tag{83}$$

$$^2\chi_{jl} = |\ldots\psi_i\bar{\psi}_i\ldots\psi_l| \tag{84}$$

The ground-state function $^2\Gamma_0$ is obtained by mixing $^2\chi_0$ with the singly excited configuration $^2\chi_{il}$, $^2\chi'_{il}$, $^2\chi_{ij}$, and $^2\chi_{jl}$. In first-order perturbation theory, ($|\lambda_{il}|$, $|\lambda'_{il}|$, $|\lambda_{ij}|$, $|\lambda_{jl}| \ll 1$), this ground-state function is expressed as

$$^2\Gamma_0 = {}^2\chi_0 + \sum_{i=1}^{j-1} \sum_{l=j+1}^{2m} \{\lambda_{il}\,{}^2\chi_{il} + \lambda'_{il}\,{}^2\chi'_{il}\} + \sum_{i=1}^{j-1} \lambda_{ij}\,{}^2\chi_{ij} + \sum_{l=j+1}^{2m} \lambda_{jl}\,{}^2\chi_{jl}\,. \tag{85}$$

The spin density in $^2\Gamma_0$ is given by the integral $[{}^2\Gamma_0 | \mathfrak{S}^{tot} | {}^2\Gamma_0]$, where $\mathfrak{S}^{tot}(=\sum_k \mathfrak{S}^{(k)})$ is the operator defined in Appendix A 1.1.

$$\begin{aligned}[{}^2\Gamma_0 | \mathfrak{S}^{tot} | {}^2\Gamma_0] &= [{}^2\chi_0 | \mathfrak{S}^{tot} | {}^2\chi_0] \\ &\quad + 2\sum_i \sum_l \{\lambda_{il}[{}^2\chi_0 | \mathfrak{S}^{tot} | {}^2\chi_{il}] + \lambda'_{il}[{}^2\chi_0 | \mathfrak{S}^{tot} | {}^2\chi'_{il}]\} \\ &\quad + 2\sum_i \lambda_{ij}[{}^2\chi_0 | \mathfrak{S}^{tot} | {}^2\chi_{ij}] + 2\sum_l \lambda_{jl}[{}^2\chi_0 | \mathfrak{S}^{tot} | {}^2\chi_{jl}]\end{aligned} \tag{86}$$

The first integral on the right is the spin density in the ground configuration $^2\chi_0$:

$$\begin{aligned}[{}^2\chi_0 | \mathfrak{S}^{tot} | {}^2\chi_0] &= \sum_{i=1}^{j-1} [\psi_i | \mathfrak{S} | \psi_i] + \sum_{i=1}^{j-1} [\bar\psi_i | \mathfrak{S} | \bar\psi_i] + [\psi_j | \mathfrak{S} | \psi_j] \\ &= \sum_{i=1}^{j-1} \psi_i^2 - \sum_{i=1}^{j-1} \psi_i^2 + \psi_j^2 = \psi_j^2\,.\end{aligned} \tag{87}$$

As in eq. (54), the contributions of electrons with paired spins cancel out, and so the spin density $\rho'(\mathbf{r})$ is given by $\psi_j^2(\mathbf{r})$, i.e. the square of a singly occupied HMO function (ψ_j is here what the orbital π was in Appendix A.1.1). The other integrals lead to the following expressions:

$$\begin{aligned}[{}^2\chi_0 | \mathfrak{S}^{tot} | {}^2\chi_{il}] &= \frac{1}{\sqrt{6}}\{[\bar\psi_i | \mathfrak{S} | \bar\psi_l] - [\psi_i | \mathfrak{S} | \psi_l]\} \\ &= \frac{1}{\sqrt{6}}\{-\psi_i\psi_l - \psi_i\psi_l\} = -\frac{2}{\sqrt{6}}\psi_i\psi_l\end{aligned} \tag{88}$$

$$\begin{aligned}[{}^2\chi_0 | \mathfrak{S}^{tot} | {}^2\chi'_{il}] &= \frac{1}{\sqrt{2}}\{[\bar\psi_i | \mathfrak{S} | \bar\psi_l] + [\psi_i | \mathfrak{S} | \psi_l]\} \\ &= \frac{1}{\sqrt{2}}\{-\psi_i\psi_l + \psi_i\psi_l\} = 0\end{aligned} \tag{89}$$

$$[{}^2\chi_0 | \mathfrak{S}^{tot} | {}^2\chi_{ij}] = [\bar\psi_i | \mathfrak{S} | \bar\psi_j] = -\psi_i\psi_j \tag{90}$$

$$[{}^2\chi_0 | \mathfrak{S}^{tot} | {}^2\chi_{jl}] = [\psi_j | \mathfrak{S} | \psi_l] = +\psi_j\psi_l\,. \tag{91}$$

Substitution of eqs. (87) - (91) into (86) yields

$$[^2\Gamma_0|\mathfrak{S}^{\text{tot}}|^2\Gamma_0] = \psi_j^2 - \frac{4}{\sqrt{6}}\sum_i\sum_l \lambda_{il}\psi_i\psi_l - 2\sum_i \lambda_{ij}\psi_i\psi_j + 2\sum_l \lambda_{jl}\psi_j\psi_l \,. \tag{92}$$

Analogously to the case of σ-π interaction, the doublet configuration $^2\chi'_{il}$ does not contribute to the spin density $\rho'(\mathbf{r})$ in the ground state $^2\Gamma_0$.

The HMO's $\psi_i(\mathbf{r})$, $\psi_j(\mathbf{r})$, and $\psi_l(\mathbf{r})$ will again be expressed as a linear combination of the AO's $\phi_\mu(\mathbf{r})$ or $\phi_\nu(\mathbf{r})$:

$$[^2\Gamma_0|\mathfrak{S}^{\text{tot}}|^2\Gamma_0] = \sum_\mu\sum_\nu \Phi_\mu\Phi_\nu\Big\{c_{j\mu}c_{j\nu} - \frac{4}{\sqrt{6}}\sum_i\sum_l \lambda_{il}c_{i\mu}c_{l\nu} - 2\sum_i \lambda_{ij}c_{i\mu}c_{j\nu} + 2\sum_l \lambda_{jl}c_{j\mu}c_{l\nu}\Big\} \,. \tag{93}$$

By integrating the right-hand side of eq. (93) over the space coordinates ($\mathbf{r}$) and using orthonormalized AO's, one obtains the following expression:

$$\sum_\mu\Big\{c_{j\mu}^2 - \frac{4}{\sqrt{6}}\sum_i\sum_l \lambda_{il}c_{i\mu}c_{l\mu} - 2\sum_i \lambda_{ij}c_{i\mu}c_{j\mu} + 2\sum_l \lambda_{jl}c_{j\mu}c_{l\mu}\Big\} \,. \tag{94}$$

The integrated spin density or the spin population ρ_μ^π at a centre μ is thus given by

$$\rho_\mu^\pi = c_{j\mu}^2 - \frac{4}{\sqrt{6}}\sum_i\sum_l \lambda_{il}c_{i\mu}c_{l\mu} - 2\sum_i \lambda_{ij}c_{i\mu}c_{j\mu} + 2\sum_l \lambda_{jl}c_{j\mu}c_{l\mu} \,. \tag{95}$$

The first term, $c_{j\mu}^2$, is the spin population at the centre μ in the ground configuration $^2\chi_0$, and invariably has a positive sign. On the other hand, the remaining terms, corresponding to the contributions of the singly excited configurations $^2\chi_{il}$, $^2\chi_{ij}$, and $^2\chi_{jl}$ to the spin population ρ_μ^π in $^2\Gamma_0$, may have either sign. The second term

$$-\frac{4}{\sqrt{6}}\sum_i\sum_l \lambda_{il}c_{i\mu}c_{l\mu} \tag{96}$$

is the only one determining the spin population ρ_μ^π, whenever the coefficient $c_{j\mu}$ at centre μ is zero. Holding the π-π spin polarization responsible for this spin population, one would expect a negative ρ_μ^π value at such a centre (cf. Section 1.6). Although there is no rigorous proof that this conclusion is correct, it is supported by many calculations, like those for the centres μ = 2 and 7 in the radical-anion of pyrene (see later).

In a first-order approximation, the coefficients $|\lambda| \ll 1$, used for the calculation of ρ_μ^π, can be equated to the cross terms between $^2\chi_0$ and the excited configurations, divided by the differences between the corresponding energies. For example, one finds for the coefficient λ_{il} the expression:

$$\lambda_{il} = -\frac{\int {}^2\chi_0 \mathscr{H}^{tot}\, {}^2\chi_{il}\, d\tau}{E_{il} - E_0} \equiv \frac{\langle {}^2\chi_0 | \mathscr{H}^{tot} | {}^2\chi_{il} \rangle}{E_{il} - E_0}, \tag{97}$$

which leads to

$$\begin{aligned} \lambda_{il} &= -\sqrt{\frac{3}{2}}\, \frac{\int \psi_i^{(1)} \psi_j^{(2)} \dfrac{e^2}{r_{12}} \psi_j^{(1)} \psi_l^{(2)}\, d\tau}{E_l - E_i} \\ &\equiv -\sqrt{\frac{3}{2}}\, \frac{\langle \psi_i \psi_j | \mathfrak{G} | \psi_j \psi_l \rangle}{E_l - E_i} \end{aligned} \tag{98}$$

Eqs. (97) and (98) are fully analogous to (67) and (73) in Appendix A 1.1. The difference between the energy E_0 of the ground configuration $^2\chi_0$ and the energy E_{il} of the excited configuration $^2\chi_{il}$ is again put approximately equal to the promotion energy $E_l - E_i$ of an electron passing from the HMO ψ_i to the HMO ψ_l. When LCAO's are substituted for the HMO's ψ_i, ψ_j, and ψ_l, the term $\langle \psi_i\, \psi_j | \mathfrak{G} | \psi_j\, \psi_l \rangle$ becomes

$$\sum_\mu \sum_\nu \sum_{\mu'} \sum_{\nu'} c_{i\mu} c_{j\nu} c_{j\mu'} c_{l\nu'} \langle \Phi_\mu \Phi_\nu | \mathfrak{G} | \Phi_{\mu'} \Phi_{\nu'} \rangle, \tag{99}$$

In the ZDO approximation, in which only the integral is taken into account

$$\gamma_{\mu\nu} = \langle \Phi_\mu \Phi_\nu | \mathfrak{G} | \Phi_\mu \Phi_\nu \rangle \equiv \int \Phi_\mu^{(1)} \Phi_\nu^{(2)} \frac{e^2}{r_{12}} \Phi_\mu^{(1)} \Phi_\nu^{(2)}\, d\tau$$

is taken into account ($\mu = \mu'$; $\nu = \nu'$), expression (99) reduces to

$$\langle \psi_i \psi_j | \mathfrak{G} | \psi_j \psi_l \rangle = \sum_\mu \sum_\nu c_{i\mu} c_{j\nu} c_{j\mu} c_{l\nu} \gamma_{\mu\nu}. \tag{100}$$

When the empirically determined values[148] of $\gamma_{\mu\nu}$ are used in eq. (100) together with the HMO coefficients, one obtains good approximations for $\langle \psi_i \psi_j | \mathfrak{G} | \psi_j \psi_l \rangle$. By using a suitable value for β, one can also conventiently estimate the energy difference $E_l - E_i$ as $(x_l - x_i)\,\beta$ from the HMO model.

This method will now be illustrated by calculaton of the spin populations ρ_μ^π in the radical-anion of pyrene. Such calculation gives negative ρ_μ^π values at centres $\mu = 2$ and $\mu = 7$[149]. The lowest antibonding HMO of pyrene (see Figure 23), i.e.

$$\begin{aligned}\psi_9 = {} & 0{,}368\,(\Phi_1 - \Phi_3 + \Phi_6 - \Phi_8) \\ & + 0{,}296\,(\Phi_4 - \Phi_5 + \Phi_9 - \Phi_{10}) - \\ & - 0{,}164\,(\Phi_{11} - \Phi_{12} + \Phi_{13} - \Phi_{14})\end{aligned} \tag{101}$$

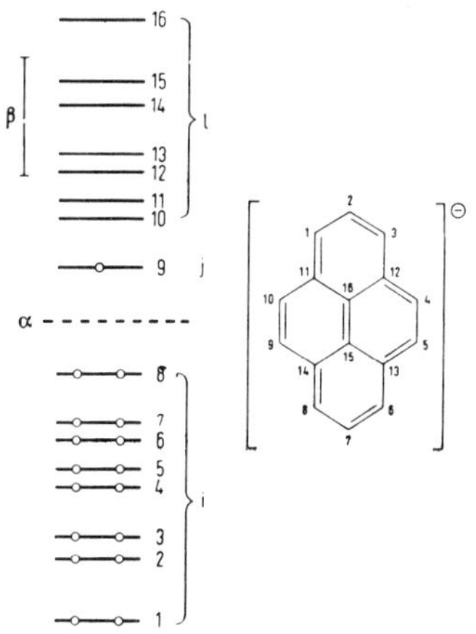

Figure 23: Energy diagram for pyrene (occupancy in the ground configuration of the radical anion).

which accommodates the unpaired electron in the ground configuration has a nodal plane traversing the centres μ = 2, 7, 15 and 16. Although the HMO values $c_{9,2}^2$ = $c_{9,7}^2$ are zero, experimentally one observes the coupling constant $a_{H2} = a_{H7} = 1.09$ gauss for the two equivalent ring protons in positions 2 and 7, this value corresponding to a spin population $\rho_2^\pi = \rho_7^\pi$ of about 0.045.

The spin population ρ_2^π (= ρ_7^π) is determined by the term (cf. eq. (96))

$$\rho_2^\pi = -\frac{4}{\sqrt{6}} \sum_i \sum_l \lambda_{il}\, c_{i2}\, c_{l2}, \tag{102}$$

where the coefficients λ_{il} are obtained by the use of eq. (98). For the sake of simplicity, only the main contributions to eq. (102), coming from the two excited configurations ${}^2\chi_{4,10}$ and ${}^2\chi_{7,13}$, will be considered. These contributions are equal, since ψ_4 and ψ_{13} on one hand, and ψ_7 and ψ_{10}, on the other, are connected by pairing properties.

$$\lambda_{4,10} = -\sqrt{\frac{3}{2}}\,\frac{\langle \psi_4 \psi_9 | \mathfrak{G} | \psi_9 \psi_{10} \rangle}{E_{10} - E_4} = \lambda_{7,13} = -\sqrt{\frac{3}{2}}\,\frac{\langle \psi_7 \psi_9 | \mathfrak{G} | \psi_9 \psi_{13} \rangle}{E_{13} - E_7}, \tag{103}$$

because

$$\langle\psi_4\psi_9|\mathfrak{G}|\psi_9\psi_{10}\rangle = \sum_\mu\sum_\nu c_{4\mu}c_{9\nu}c_{9\mu}c_{10\nu}\gamma_{\mu\nu}$$
$$= \langle\psi_7\psi_9|\mathfrak{G}|\psi_9\psi_{13}\rangle = \sum_\mu\sum_\nu c_{7\mu}c_{9\nu}c_{9\mu}c_{13\nu}\gamma_{\mu\nu} \qquad (104)$$

and

$$E_{10}-E_4 = (x_{10}-x_4)\beta = E_{13}-E_7 = (x_{13}-x_7)\beta. \qquad (105)$$

Furthermore,

$$c_{4,2}\times c_{10,2} = c_{7,2}\times c_{13,2}. \qquad (106)$$

Using the HMO coefficients $c_{i\mu}$ and $c_{l\mu}$ as well as the Pariser-Parr [148] parameters $\gamma_{\mu\nu} = \langle\phi_\mu\phi_\nu|\mathfrak{G}|\phi_\mu\phi_\nu\rangle$, Hoijtink et al.[149] obtained +0.246 eV for the cross term (104). The estimated value of β is –2.5 eV, so that the energy difference (105), which is $-2.227\,\beta$ in the HMO model, has the value 5.57 eV. This gives $\lambda_{4,10} = \lambda_{7,13} = -0.054$, substitution of which into eq. (102) together with

$$c_{4,2}\times c_{10,2} = c_{7,2}\times c_{13,2} = -0.196$$

leads to the result $\rho_2^\pi = \rho_7^\pi = -\ 0.035$. One may note that the corresponding value obtained with the aid of McLachlan's formula (eq. (44)) is – 0.054 (cf. Table 2, Section 2.1).

The pyrene radical-anion has played an important role in this connection, partly by being the first radical-ion whose hyperfine structure led to the postulation of negative π-spin populations[145,149] (previouly this had been done only for the neutral, or odd, radicals, triphenylmethyl[139] and phenalenyl[151]), and partly because it was this species whose proton resonance spectrum taken in solid phase at 4 °K showed that the coupling constant a_{H2} differed in sign from the other two coupling constants a_{H1} and a_{H4}[150], in agreement with the assumption of a negative spin population ρ_2^π.

The calculation of the spin populations ρ_μ^π in the radical-anion of cycl [3,2,2] azine (cf. Section 2.2) is another example of the application of the CI method in this field.[88a]

A. 1.3. Anisotropy effects

It was mentioned in Section 1.1 that the anisotropic hyperfine interaction $(\delta E)_{aniso}$ contributes to the width $\Delta\nu = \gamma_E\ \Delta H$ of the ESR lines of radicals in solution. A further contribution to the line-width comes from the anisotropic nature of the g_E factor*), for the latter is not fully averaged out (as is the hyperfine anisotropy) by the molecular tumbling motion. The total contribution to the line-width due to anisotropic effects can be resolved into three components:

$$(\Delta\nu)_{aniso} = (\Delta\nu)^{gg}_{aniso} + (\Delta\nu)^{ga}_{aniso} + (\Delta\nu)^{aa}_{aniso} \tag{107}$$

Instead of their exact mathematical formulation, which requires the use of tensors, it is sufficient for the present purposes to say that $(\Delta\nu)^{gg}_{aniso}$, and $(\Delta\nu)^{aa}_{aniso}$ depend on the square of the anisotropic g_E tensor and on the square of the anisotropic hyperfine tensor, respectively,while the cross term $(\Delta\nu)^{ga}_{aniso}$ is a function of the product of the two tensors[302,308]. The anisotropic hyperfine tensor describes the direction-dependence of the anisotropic hyperfine interaction $(\delta E)_{aniso}$, i.e. the dipole-dipole interaction between the unpaired electron and a magnetic nucleus X (cf. Section 1.1). Like the Fermi contact term $(\delta E)_{iso}$ specified in eq. (12), this interaction varies linearly with the spin quantum number M_I (X) of nucleus X.

The terms $(\Delta\nu)^{ga}_{aniso}$ and $(\Delta\nu)^{aa}_{aniso}$ therefore include the quantum number $M_I(X)$ having the same power as the anisotropic hyperfine tensor. The variation of $(\Delta\nu)_{aniso}$ with the quantum number $M_I(X)$ can thus be expressed as follows:

$$(\Delta\nu)_{aniso} = A + BM_I(X) + CM_I^2(X). \tag{108}$$

The fact that $(\Delta\nu)_{aniso}$ is a function of $M_I(X)$ has far-reaching consequences, namely, that the individual hyperfine lines of one and the same ESR spectrum can differ in width.

Fig. 2 in Section 1.1 shows that there is a unique connection between the hyperfine line of an ESR spectrum and the value of the spin quantum number $M_I(X)$ of nucleus X, since $M_I(X)$ characterizes the two sub-levels which are involved in the transition and for which selection rule (13) gives the same nuclear spin state. The magnitude of $M_I(X)$ is given by the position of the line in the spectrum. The sign of $M_I(X)$ is related to the sign of the coupling constant a_X of nucleus X. In the case of a positive constant a_X, the lines with positive $M_I(X)$ and those with negative $M_I(X)$ are situated in the low-field and the high-field region of the spectrum, respectively. The reverse situation occurs when the coupling constant is negative.

*) The g_E anisotropy has not been mentioned so far, since it is insignificant in the case of organic radicals. This is caused by the same circumstances that are responsible for the fact that the g_E factors of organic radicals differ only negligibly from the g_E factor of free electrons (absence of orbital paramagnetism and only weak spin-orbital coupling; cf. Section 1.1).

All hyperfine lines in an ESR spectrum resulting from the interaction of the unpaired electron with nucleus X in a given radical will have the same values for A, B, and C in eq. (108). While A and C are positive, the sign of B depends on the signs of the following three quantities[111,308]: (i) the g_N value of the nucleus X; (ii) the spin population at this nucleus (local spin population); and (iii) the difference $2g_E^z - (g_E^x + g_E^y)$, where g_E^x, g_E^y, and g_E^z are the three principal components of the g_E tensor, the z direction being taken perpendicular to the aromatic xy plane. Since the theory stipulates a negative sign for the difference $2g_E^z - (g_E^x + g_E^y)$, B is negative if g_N and ρ_μ^π have the same sign, and positive if they have opposite signs. The term $(\Delta\nu)_{\text{aniso}}^{gg}$ (=A) makes an equal positive contribution to the width of all hyperfine lines. The term $(\Delta\nu)_{\text{aniso}}^{aa}$ (= $CM_I^2(X)$) likewise makes a positive contribution. The effect of this term is that the lines toward the ends of the spectrum ($|M_I(X)|$ large) are broader than those near the centre of the spectrum ($|M_I(X)| = 0$ or small). By contrast, the term $(\Delta\nu)_{\text{aniso}}^{ga}$ (=B $M_I(X)$) can have either sign; for $B < 0$, it is positive or negative according to whether $M_I(X)$ is negative or positive. In the case of $B < 0$, the connection between the sign of the coupling constant a_X and that of the spin quantum number $M_I(X)$ therefore specifies that: a_X is positive, if the lines are broader in the high-field region; and a_X is negative, if the lines are broader in the low-field region. The opposite would be true for $B > 0$. When the sign of B is known, the observation that the hyperfine lines have different widths on the two sides of the spectrum therefore permits the sign or the coupling constant a_X to be determined.

It has already been mentioned that the less completely the hyperfine anisotropy is averaged out by the molecular motion, the more it contributes to the line width. For this reason, such anisotropy, is more evident in viscous solvents than in non-viscous ones (cf. Section 1.1 and 1.3 together with Fig. 9). The same applies to the g_E anisotropy. Furthermore, in the case of anisotropic hyperfine interaction, the distance between the two dipoles involved is expected to assume a decisive role (the dipole-dipole interaction varies as the inverse cube of this distance). A small 'mean' distance between the unpaired electron and the nucleus X is thus a prerequisite for an appreciable hyperfine anisotropy, which is fulfilled when the π-spin population is high near nucleus X (local spin population). In aromatic radicals, the ^{13}C and the ^{14}N nuclei are situated in the π-electron centres themselves, so that, provided the spin population ρ_μ^π is high at these centres μ, the hyperfine anisotropy will be large. For ring protons and especially for the protons of alkyl substituents, on the other hand, the anisotropic effects are considerably smaller. The method of determining the sign of a_X from the variation of the line broadening $(\Delta\nu)_{\text{aniso}}$ with $M_I(X)$ has therefore only limited applicability; it frequently works when X = ^{13}C and ^{14}N, in particular, if the local spin population ρ_μ^π is high, but it cannot generally be used when X = ^1H. For-

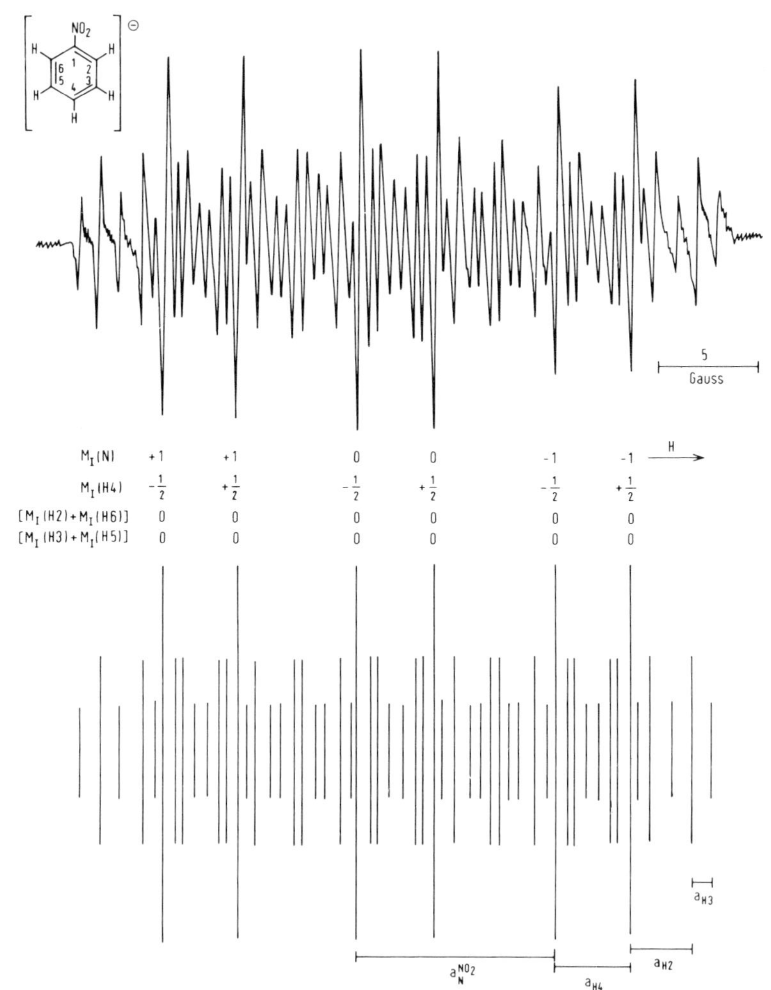

Figure 24: ESR spectrum of the radical-anion of nitrobenzene (in N,N-dimethylformamide, at + 25 °C; gegenion tetraethylammonium$^{\oplus}$) and the reconstructed hyperfine structure (stick diagram).

tunately, the case of ^{13}C and ^{14}N nuclei in centres μ of large ρ_{μ}^{π} values is just that for which the sign of B can be assumed negative with a high degree of certainty. This is so because the g_N values of the two nuclei are positive (cf. Table 1), and because usually high π-spin populations have positive signs as well (cf. Section 1.5).

When the hyperfine structure results not from a single nucleus, but from a set of n equivalent nuclei X_i having the same coupling constant a_X the spin quantum numbers must be summed for the whole set (cf. Figure 3), so that eq. (108) is replaced by

$$(\Delta\nu)_{aniso} = A + B\sum_{i=1}^{n} M_I(X_i) + C[\sum_{i=1}^{n} M_I(X_i)]^2. \tag{109}$$

An aromatic radical-ion usually contains more than one set of equivalent magnetic nuclei, and thus more than one $\sum_i M_I(X_i)$ value is required to denote the hyperfine lines. Eq. (109) is visibly complicated when several sets make an appreciable contribution to $(\Delta\nu)_{aniso}$, but it is often sufficient to consider those nuclei for which the hyperfine anisotropy has a measurable value; for all other nuclei, the anisotropic effect may be neglected. Eqs. (108) and (109) can therefore be used with aromatic radical-ions containing, in addition to protons, only one set of such equivalent nuclei (^{13}C or ^{14}N) in π-electron centres of high spin population. This will be illustrated in the case of the nitrobenzene radical-anion, whose ESR spectrum is shown in Figure 24. In this case, there is a high π-spin population at the nitrogen, and the ^{14}N nucleus gives rise to the main splitting, denoted by the coupling constant $a_N^{NO_2}$ (cf. Section 2.4). The other magnetic nuclei are ring protons, namely one proton in position 4 and two pairs of equivalent protons, in positions 2, 6 and 3, 5, respectively, the corresponding coupling constants being a_{H4}, $a_{H2} = a_{H6}$, and $a_{H3} = a_{H5}$. The total number of hyperfine lines is therefore $3 \times 2 \times 3^2 = 54$. The six strongest lines appear in three pairs, one in the low-field region, one in the middle, and one in the high-field region of the spectrum. The distance between the two components of a line pair yields the coupling constant a_{H4}, while the much larger distance between two corresponding components of unlike line pairs gives the coupling constant $a_N^{NO_2}$. The peak height is the same for the twin components of a pair, but it varies from pair to pair. Since according to the reconstructed spectrum (stick diagram), the same intensity is to be expected for all six lines, the difference in height must be due to the different widths of the three line pairs*).

The four spin quantum numbers belonging to the ^{14}N nucleus, the ring proton in position 4, and the two pairs of equivalent ring proton in positions 2,6 and 3,5 are denoted by $M_I(N)$, $M_I(H4)$, $[M_I(H2) + M_I(H6)]$ and $[M_I(H3) + M_I(H5)]$, respectively.

*) The peak height is a measure of the intensity only when the lines have the same width, because it is the area under the spectral signal (integrated intensity) that is a true measure of the intensity. Therefore, when the signals have the same integrated intensity, the heights are inversely proportional to the widths. In the case of ESR hyperfine lines (which are recorded as the derivative dA/dH of the absorption curve with respect to the field strength), the observed line-height varies as the inverse square of the line-width $\Delta\nu = \gamma_E \Delta_H$. For example, a twofold broadening reduces the height to a quarter.

The reconstructed spectrum indicates that the last two values, $[M_I(H2) + M_I(H6)]$ and $[M_I(H3) + M_I(H5)]$, are zero for the six line pairs involved. The quantum number $M_I(H4)$ can be either $+\frac{1}{2}$ or $-\frac{1}{2}$, so that the two components of a line pair separated by a_{H4} will differ in the sign of $M_I(H4)$. The assignment of $+\frac{1}{2}$ to the high-field component and $-\frac{1}{2}$ to the low-field component of the line pair (Figure 24) is based on theoretical considerations. McConnell's formula (20) gives namely a negative sign for a_{H4}, since according to the MO model the spin population ρ_4^π is positive at the centre 4.

The spin quantum number $M_I(N)$, which must be different for the line pairs (distance $a_N^{NO_2}$), has the values +1,0 and −1. The zero value can be assigned unambiguously to the middle line pair, while the assignment of the +1 and the −1 value of $M_I(N)$ to the line pairs at high-field and low-field, respectively, rests on a comparison of the line-widths. A quick glance at Figure 24 is sufficient to make one realize that the differences in the widths of the three line pairs are related to the spin quantum numbers, $M_I(N)$, and they must therefore be due to anisotropic hyperfine interactions with the ^{14}N nucleus. The same conclusion is reached by considering that the anisotropic effects cannot be due to the ring protons, while the ^{14}N nucleus, having a high local π-spin population, fulfils all the conditions for causing such effects. The differences in the widths of the three line pairs can therefore be expressed as follows:

$$(\Delta\nu)_{aniso} = A + B\,M_I(N) + C\,M_I^2(N) \qquad (110)$$

For reasons mentioned before (positive signs of the g_N value of the ^{14}N nucleus and of the high local spin population at the nitrogen), the sign of B in eq. (110) is negative.

The line pair in the middle ($M_I(N) = 0$) is higher than the outside ones ($M_N(N) = \pm 1$), and particularly higher than that at the high-field. Since a decrease in line-height results from an increase in the line-width $\Delta\nu$, the following sequence must be true:

$$\Delta\nu, \text{middle} < \Delta\nu, \text{low-field} < \Delta\nu, \text{high-field}$$

Eq. (110) with A and C > 0, and B < 0 gives the correct relationship $\Delta\nu$, low-field < $\Delta\nu$, high-field, when a positive and a negative spin quantum number $M_I(N)$ are assigned to the line pairs in low-field and high-field regions, respectively. One thus obtains for $M_I(N)$ and $(\Delta\nu)_{aniso}$:

middle	: $M_I(N) = 0$,	$(\Delta\nu)_{aniso} = A$
low-field	: $M_I(N) = +1$,	$(\Delta\nu)_{aniso} = A + B + C$
high-field	: $M_I(N) = -1$,	$(\Delta\nu)_{aniso} = A - B + C$

The above assignment means that the coupling constant $a_N^{NO_2}$ must be positive as suggested by other considerations (cf. Section 2.4). Furthermore, the relationship $\Delta\nu$, middle $<$ $\Delta\nu$, low-field indicates that $C| > |B|$ in the present case.

2. Individual Systems

Preliminary Remarks

The following symbols, most of which were introduced in Part 1, will be used here:

μ	π-electron centre ($\tilde{\mu}$ denotes a substituted centre)
$a_{H\mu}$	coupling constant of a proton linked to centre μ (generally a ring proton)
$a_H^{CH_n}$	coupling constant of an alkyl proton (n = 2 and 3 for CH_2 and CH_3 respectively)
$a_{N\mu}$	coupling constant of a ^{14}N nucleus at centre μ in a ring
a_N^X	coupling constant of a ^{14}N nucleus in substituent X (X = CN, NO_2 , etc.)
a_F	coupling constant of a ^{19}F nucleus
$a_{C\mu}$	coupling constant of a ^{13}C nucleus in centre μ of a ring
a_C^X	coupling constant of a ^{13}C nucleus in substituent X (X = CH_n, CN, etc.)
a_{Me}	coupling constant of an alkali metal nucleus (Me = ^7Li, ^{23}Na, ^{39}K, etc.)
ρ_μ^π	π-spin population at centre μ
$\rho_{H_3}^{1s}$	ls-spin population at pseudo-centre H_3 of a methyl substituent
ρ_{Me}^{ns}	ns-spin population at an alkali metal cation (n = 2, 3, 4 for Li, Na, and K, respectively)
$c_{j\mu}^2$	HMO spin population (square of the LCAO coefficient at centre μ for the singly occupied HMO ψ_j); in particular:
$c_{a\mu}^2$	HMO spin population when $\psi_j \equiv \psi_a$ = lowest vacant HMO of the neutral system (generally an antibonding orbital)

$c^2_{b\mu}$ HMO spin population when $\psi_j \equiv \psi_b$ = highest occupied HMO of the neutral system (generally a bonding orbital)

$c^2_{n\mu}$ HMO spin population when $\psi_j \equiv \psi_n$ = non-bonding HMO (used only for th non-bonding orbitals of the 4r-perimeters; cf. Section 2.5)

Superscripts ⊖ and ⊕ denote radical-anions and radical-cations, respectively. All the coupling constants are given in gauss.

In the case of some of the coupling constants specified below, the assignment to the various sets of equivalent nuclei is based on comparison with the calculated π-spin populations, and - as pointed out in Section 1.4 - such a comparison permits a reliable assignment of only those coupling constants which differ greatly in magnitude. For coupling constants of similar magnitude, the assignment may depend on the choice of the approximation method used to calculate the spin populations. In the following discussion, the assignments proposed by the various authors will be reproduced even when they are supported only by a special MO model. This often makes for poorer agreement of experimental data with spin populations obtained by another model.

2.1. Hydrocarbons

In the case of unsubstituted systems containing no heteroatoms, the calculation of the π-spin population by the approximation methods described in Section 1.6 requires the smallest number of empirical parameters, so that the resulting theoretical values ρ^{π}_{μ} may be considered as relatively reliable. The radical-ions of unsubstituted hydrocarbons are therefore used with advantage to check the validity of the relationship

$$a_{H_\mu} = Q_{CH}\rho^{\pi}_{\mu} \cdot \tag{20}$$

Alternant systems. Table 2 shows the coupling constants $a_{H\mu}$ of the ring protons and the HMO spin populations $c^2_{j\mu}$ at the carbon centres μ, for the radical-ions of some alternant hydrocarbons. Since the HMO's of these systems exhibit pairing properties (cf. Section 1.6), the anion and the cation of any given compound have the same values $c^2_{j\mu}$ (i.e. $c^2_{a\mu} = c^2_{b\mu}$).

The regression of $a_{H\mu}$ on $c^2_{j\mu}$ is shown in Figure 25. The standard deviation of individual measurements from the regression line with the equation

$$a_{H_\mu} = \cdot 0.16 + 29.0\, c^2_{j\mu} \tag{111}$$

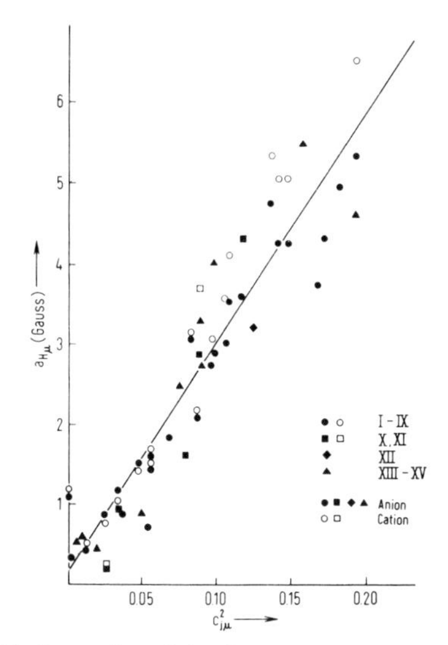

Figure 25: Regression of the ring proton coupling constants $a_{H\mu}$ on the HMO spin populations $c^2_{j\mu}$ for alternant hydrocarbon radical-ions listed in Table 2.

is 0.58 gauss. The correlation between the observed and the calculated values is good, particularly in view of the fact that such widely varying systems as cata- or peri-condensed acenes (compounds I-XI), compounds with four- and eight-membered rings*) (X-XII), and hydrocarbons formed from weakly coupled benzene and ethylene residues (XIII-XV) are included. This should be pointed out because many of the molecules mentioned contain relatively long C-C bonds, and in such cases the use of different values of the HMO parameters β may lead to better agreement between theory and experiment. However, in order not to impair the validity of the test by additional parameters, any variation in β, both for the alternant and the non-alternant hydrocarbon radical-ions (see Table 3 below), was precluded.

The regression line does not deviate significantly from the origin of the coordinate system. Its slope, 29.0 ± 1.4 gauss, is an empirical estimate for the parameter $|Q_{CH}|$ in eq. (20). As mentioned in Section 1.5 and Appendix A. 1.1, the sign of this parameter is negative, but it cannot be determined from ESR spectra of radicals in solution. It is clear that the empirical estimate of $|Q_{CH}|$ depends on the method

*)Unlike the neutral compounds XII, the radical-anion $XII^{\ominus}$ of cyclooctatetraene is considered to be planar (cf. Section 2.5)

used to calculate the π-spin populations. The experimental data listed in Table 2 yield a smaller magnitude for this parameter, when the HMO quantities $c_{j\mu}^2$ are replaced by more exact values. These more exact spin populations ρ_μ^π may also be negative, particularly those at centres μ with vanishing HMO quantities $c_{j\mu}^2$ (cf. Section 1.6 and Appendix A 1.2). In order that the condition (21) for the sum of ρ_μ^π remains valid, the apperance of such negative spin populations must be accompanied by a compensating increase in the positive values at other centres. The more exact values $|\rho_\mu^\pi|$ are therefore generally larger than the HMO quantities $c_{j\mu}^2$. This can be seen from Table 2, which also contains the spin populations ρ_μ^π that have been calculated from $c_{j\mu}^2$ with the aid of McLachlan's equation [eq. (44)]. Like the HMO values $c_{j\mu}^2$, the resulting spin populations ρ_μ^π are equal for the radical-anion and the radical-cation of any given alternant hydrocarbon, i.e. $\rho_\mu^{\pi\ominus} = \rho_\mu^{\pi\oplus}$.

Table 2: Coupling constants of ring protons and spin populations in the radical-ions of some alternant hydrocarbons.

Radical ion of	μ	$a_{H\mu}^{\ominus}$	$a_{H\mu}^{\oplus}$	$c_{j\mu}^2$ ($c_{a\mu}^2 = c_{b\mu}^2$)	ρ_μ^π ($\rho_\mu^{\pi\ominus} = \rho_\mu^{\pi\oplus}$)	Ref.
I Benzene	1	3,75		0,167	0,167	36)
II Naphthalene	1	4,95		0,181	0,229	82)
	2	1,83		0,069	0,043	
III Anthracene	1	2,74	3,06	0,097	0,119	
	2	1,51	1,38	0,048	0,031	153)
	9	5,34	6,53	0.193	0.259	
IV Tetracene	1	1,55	1,69	0,056	0,067	
	2	1,15	1,03	0,034	0,021	154,90)
	5	4,25	5,06	0,147	0,197	

Radical ion of	μ	$a_{H\mu}^{\ominus}$	$a_{H\mu}^{\oplus}$	$c_{j\mu}^2$ ($c_{a\mu}^2 = c_{b\mu}^2$)	ρ_μ^π ($\rho_\mu^{\pi\ominus} = \rho_\mu^{\pi\oplus}$)	Ref.
V Pentacene	1	0,92	0,98	0,035	0,041	
	2	0,87	0,76	0,025	0,015	45,153)
	5	3,03	3,56	0,106	0,140	
	6	4,26	5,08	0,141	0,190	
VI Phenanthrene	1	3,60		0,116	0,160	
	2	0,32		0,002	−0,044	
	3	2,88		0,099	0,126	93)
	4	0,72		0,054	0,038	
	9	4,32		0,172	0,198	
VII Pyrene	1	4,75	5,38	0,136	0,189	
	2	1,09	1,18	0	−0,054	149,59)
	4	2,08	2,12	0,087	0,093	
VIII Perylene	1	3,08	3,10	0,083	0,107	
	2	0,46	0,46	0,013	−0,023	154)
	3	3,53	4,10	0,108	0,153	
IX Coronene		1,47	1,53	0,056	0,061	45,46)
X Biphenylene	1	0,21	0,21	0,027	−0,009	48)
	2	2,86	3,69	0,087	0,091	

Radical ion of	μ	$a^{\ominus}_{H\mu}$	$a^{\oplus}_{H\mu}$	$c^2_{j\mu}$ ($c^2_{a\mu} = c^2_{b\mu}$)	ρ^{π}_{μ} ($\rho^{\pi\oplus}_{\mu} = \rho^{\pi\oplus}_{\mu}$)	Ref.
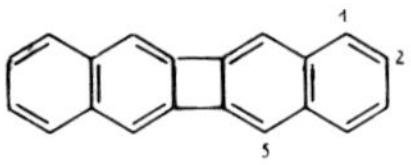 XI Binaphthylene	1	1,62		0,079	0,102	
	2	0,93		0,034	0,024	48)
	5	4,31		0,117	0,158	
 XII Cycloocta-tetraene		3,21		0,125	0,125	155)
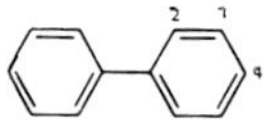 XIII Biphenyl	2	2,73		0,090	0,106	
	3	0,43		0,020	−0,026	32)
	4	5,46		0,158	0,213	
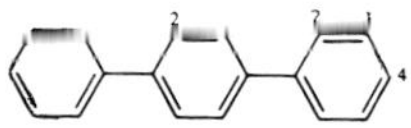 XIV Terphenyl	2	2,07		0,060	0,078	
	3	0,56		0,008	−0,023	
	4	3,30		0,088	0,122	156)
	2'	0,95		0,049	0,037	
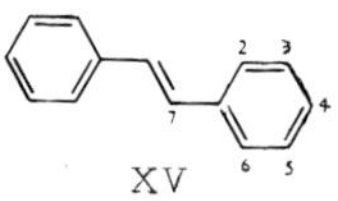 XV Stilbene	2	3,03 } 2,48 average		0,075	0,095	
	6	1,94				
	3	0,83 } 0,56 average		0,006	−0,032	157)
	4	0,30				
	5	4,00		0,098	0,132	
	7	4,51		0,192	0,211	

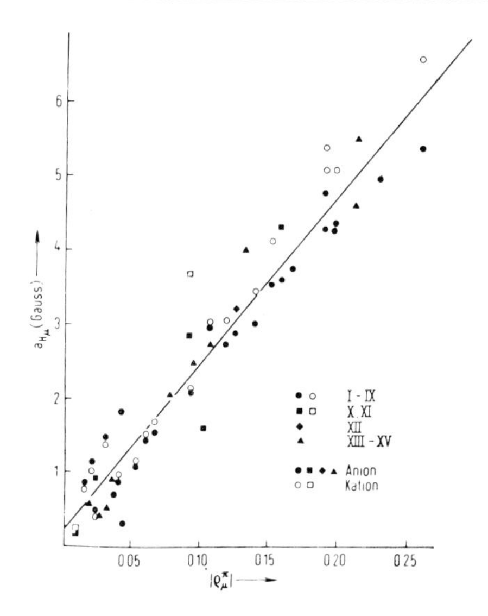

Figure 26: Regression of the ring proton coupling constants $a_{H\mu}$ on the spin populations $|\rho_\mu^\pi|$ calculated with McLachlan's formula (44) for alternant hydrocarbon radical-ions listed in Table 2.

The equation of the regression line shown in Figure 26 is

$$a_{H\mu} = 0.22 + 22.9\,|\rho_\mu^\pi|. \tag{112}$$

This equation gives a new and more realistic estimate of $|Q_{CH}| = 22.9 \pm 0.8$ gauss. The standard deviation of individual measurements from the regression line amounts now to 0.44 gauss. This result means that refinement of the model improves somewhat the correlation between the calculated and the observed values. The regression line is again close to the origin of the coordinate system.

Table 2 and Figures 25 and 26 show that the coupling constants $a_{H\mu}^{\ominus}$ and $a_{H\mu}^{\oplus}$ of the radical-ions derived from the same parent compound differ only to a small extent. This relationship is fully reflected in the similarities between the ESR spectra of the radical-anions and those of corresponding cations, as illustrated on the example of the perylene radical-ions [44] in Figure 27.

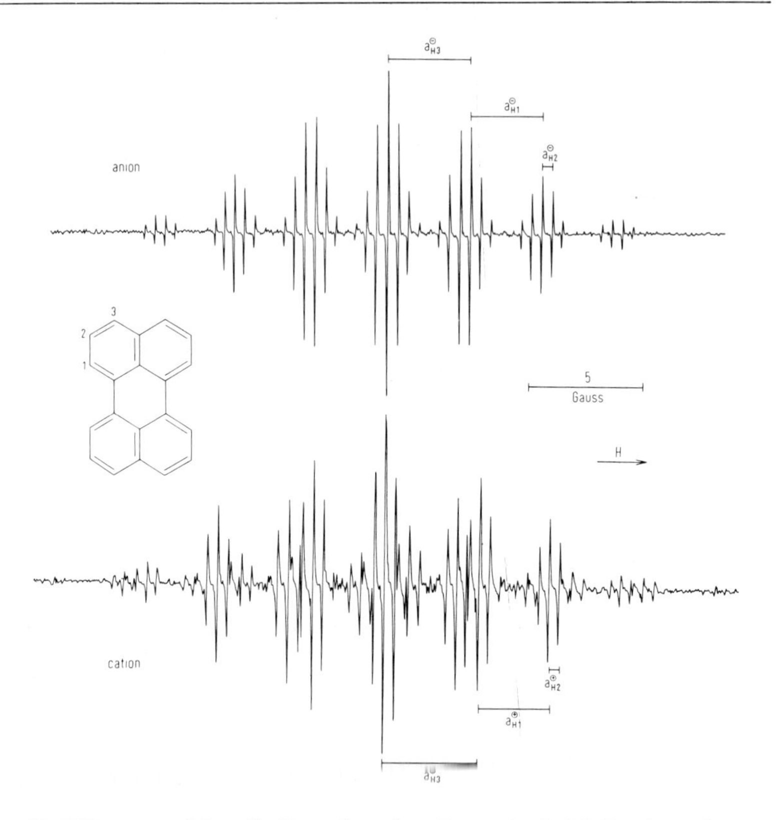

Figure 27: ESR spectra of the radical-ions of perylene. Top: anion in 1,2-dimethoxyethane, at −70 °C; gegenion $Na^{\oplus}$; bottom: cation in conc. sulfuric acid, at +25 °C.
(Due to the accidental degeneracy, $a_{H1}^{\ominus} + a_{H2}^{\ominus} \approx a_{H3}^{\ominus}$, the spectrum of the anion apparently contains fewer lines than that of the cation).

The experimental results thus support the MO prediction that in the case of the alternant radical-ions, the anion and the cation of the same hydrocarbon should have equal spin populations, i.e. $c_{a\mu}^2 = c_{b\mu}^2$ and $\rho_{\mu}^{\pi\ominus} = \rho_{\mu}^{\pi\oplus}$.

Starting from equal spin populations $\rho_{\mu}^{\pi\ominus}$: $\rho_{\mu}^{\pi\oplus}$, one requires at least two parameters Q to reproduce the observed small differences in the coupling constants $a_{H\mu}^{\ominus}$ and $a_{H\mu}^{\oplus}$ of the corresponding alternant radical-ions.

The simplest method capable of giving more or less correct results is based on the fact that the coupling constants $a_{H\mu}^{\oplus}$ of the radical-cations are in general slightly larger than the coupling constants $a_{H\mu}^{\ominus}$ of the corresponding anions. The agreement between experiment and theory can therefore be improved by the use of somewhat different $|Q_{CH}|$ values for cations and anions, i.e. $|Q_{CH}^{\oplus}|$ and $|Q_{CH}^{\ominus}|$:

$$a_{H\mu}^{\ominus} = Q_{CH}^{\ominus}\rho_{\mu}^{\pi} \tag{113}$$

$$a_{H\mu}^{\oplus} = Q_{CH}^{\oplus}\rho_{\mu}^{\pi}\,. \tag{114}$$

The following Q_{CH}-values have been obtained from the data listed in Table 2:

	$\lvert Q_{CH}^{\ominus}\rvert$	$\lvert Q_{CH}^{\oplus}\rvert$
Calculated with HMO quantities $c_{j\mu}^2$	26.9 gauss	35.1 gauss
Calculated with more accurate McLachlan spin populations ρ_{μ}^{π}	21.8 gauss	25.4 gauss.

In other methods, a second parameter Q_2 is introduced in addition to Q_1 ($=Q_{CH}$). Formulae (115) and (116) are then used. The equation [154,158]

$$a_{H\mu} = (Q_1 \pm Q_2\,\varepsilon_{\mu})\,\rho_{\mu}^{\pi} \tag{115}$$

takes into account the excess charge ϵ_{μ}^{π} ($=|q_{\mu}^{\pi}-1|$) at the centre μ, the positive and the negative signs referring to radical-cations and radical-anions. By substituting $c_{j\mu}^2$ for ρ_{μ}^{π} as an approximation, and considering that in an alternant system the HMO value of the excess charge ϵ_{μ}^{π} is also $c_{j\mu}^2$, one can simplify eq. (115) to

$$a_{H\mu} = Q_1\,c_{j\mu}^2 \pm Q_2\,c_{j\mu}^4\,. \tag{116}$$

The empirical values of $|Q_1|$ and $|Q_2|$ are respectively [158] 27 and 12 gauss. The equation [159]

$$a_{H\mu} = Q_1\,c_{j\mu}^2 \pm Q_2 \sum_{\nu} |c_{j\mu}c_{j\nu}| \tag{117}$$

contains a correction term obtained from eqs. (74) and (75) in Appendix A.1.1 by retention of the products of the HMO coefficients $c_{j\mu}$ and $c_{j\nu}$ at the adjacent centres μ

and ν. The positive and the negative sign again refer to radical-cations and radical-anions. The empirical values of $|Q_1|$ and $|Q_2|$ are this time 32 and 7 gauss, respectively. A further possibility is a combination of eqs. (116) and (117), leading to an expression containing three parameters[160]:

$$a_{H\mu} = Q_1 c_{j\mu}^2 \pm Q_2 \sum_{\nu} |c_{j\mu} c_{j\nu}| \pm Q_3 c_{j\mu}^4 . \qquad (118)$$

The following values have been proposed for these parameters: $|Q_1| = 32$, $|Q_2| = 2$, and $|Q_3| = 10$ gauss[299*].

Non-alternant systems. There are only a few non-alternant hydrocarbon radical-ions known[49,55,67,161]. The coupling constants $a_{H\mu}$ of their ring protons are listed in Table 3, together with the corresponding spin populations at the carbon centres μ.

Table 3: Coupling constants of ring protons and spin populations in the radical-ions of some non-alternant hydrocarbons.

Radical ion of	μ	$a_{H\mu}^{\ominus}$	$a_{H\mu}^{\oplus}$	$c_{a\mu}^2$	$c_{b\mu}^2$	$\rho_{\mu}^{\pi\ominus}$	$\rho_{\mu}^{\pi\oplus}$	Ref.
XVI Azulene	1	0,27		0,004		−0,025		67)
	2	3,95		0,100		0,109		
	4	6,22		0,221		0,304		
	5	1,34		0,011		−0,085		
	6	8,83		0,261		0,375		
XVII Acenaphthylene	1	3,09		0,104		0,100		162)
	3	4,51		0,151		0,205		
	4	0,45		0,014		−0,052		
	5	5,64		0,178		0,261		

*) It has recently been pointed out[293] that the differences between coupling constants $a_{H\mu}^{\ominus}$ and $a_{H\mu}^{\oplus}$ of the two corresponding radical-ions may also be due to causes other than charge differences[154,158] and perturbation by adjacent centres[159]. In this connection it would seem profitable to investigate particularly the effect of solvents, which are different for radical-anions and radical-cations (cf. Section 1.2).

Radical ion of	μ	$a_{H\mu}^{\ominus}$	$a_{H\mu}^{\oplus}$	$c_{a\mu}^2$	$c_{b\mu}^2$	$\rho_\mu^{\pi\ominus}$	$\rho_\mu^{\pi\oplus}$	Ref.
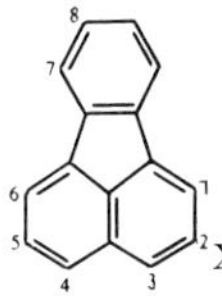 XVIII	1	3,90		0,121		0,164		130)
	2	0,17		0,022		−0,032		
	3	5,20		0,163		0,241		
	7	0,08		0,016		−0,037		
Fluoranthene	8	1,21		0,039		0,037		
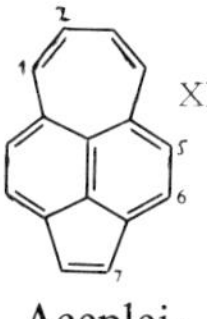 XIX	1	0,21	4,53	0,027	0,116	−0,003	0,147	49)
	2	2,76	2,13	0,087	0,056	0,108	0,040	
	5	0,80	5,88	0,000	0,143	−0,057	0,206	
	6	4,04	0,78	0,136	0,007	0,195	−0,045	
Aceplei-adylene	7	2,44	2,70	0,087	0,056	0,096	0,050	
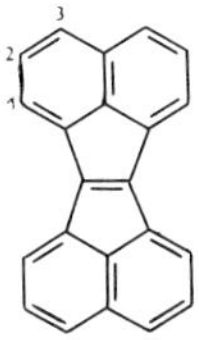 XX	1	3,30	1,00	0,093	0,030	0,132	0,035	49)
	2	0,71	0,24	0,003	0,001	−0,038	−0,021	
Acenaphth-[1,2-a]ace-naphthylene	3	3,35	1,76	0,100	0,035	0,149	0,045	

Since the HMO's ψ_a and ψ_b of non-alternant systems do not have the pairing properties mentioned earlier, the HMO spin populations are in general different for the radical-anion and the radical-cation of a given hydrocarbon ($c_{a\mu}^2 \neq c_{b\mu}^2$). The same applies to the spin populations calculated from these HMO-values with the aid of McLachlan's equation (44), i.e. $\rho_\mu^{\pi\ominus} \neq \rho_\mu^{\pi\oplus}$.

The data in Table 3 show that there is a good agreement between the measured and calculated values. This agreement is particularly impressive in the case of acepleiadylene (XIX)[271] and acenapht[1,2-a]acenaphthylene (XX)[272] which yield both radical-anions and radical-cations. It is illustrated by the following comparison between the summed absolute magnitudes of the calculated spin populations at proton-carrying centres μ and the overall ranges of the spectra (cf. eq. (18), Section 1.5).

Radical-anion	$\sum_\mu c_{a\mu}^2$	$\sum_\mu \lvert\rho_\mu^{\pi\ominus}\rvert$	$\sum_\mu \lvert a_{H\mu}^{\ominus}\rvert$	Radical-cation	$\sum_\mu c_{b\mu}^2$	$\sum_\mu \lvert\rho_\mu^{\pi\oplus}\rvert$	$\sum_\mu \lvert a_{H\mu}^{\oplus}\rvert$
$XIX^{\ominus}$	0.674	0.918	20.49	$XIX^{\oplus}$	0.756	0.976	32.04
$XX^{\ominus}$	0.784	1.276	29.44	$XX^{\oplus}$	0.268	0.404	12.00

The ratio $\Sigma_\mu \lvert a_{H\mu}^{\oplus}\rvert / \Sigma_\mu \lvert a_{H\mu}^{\ominus}\rvert$ of the overall spectral ranges is 1.56 for the radical-ions of XIX and 0.41 for those of XX. On the other hand, in the case of the radical-ions of alternant hydrocarbons, the corresponding value is always about 1.1. Thus the consequence of the pairing properties of the HMO's is reflected in the values of this ratio, some of which are listed in Table 4.

Table 4: Overall spectral ranges of some hydrocarbon radical-ions.

Radical ion of	$\sum_\mu \lvert a_{H\mu}^{\oplus}\rvert$	$\sum_\mu \lvert a_{H\mu}^{\ominus}\rvert$	$\dfrac{\sum_\mu \lvert a_{H\mu}^{\oplus}\rvert}{\sum_\mu \lvert a_{H\mu}^{\ominus}\rvert}$	Ref.
alternant:				
Anthracene III	30,82	27,68	1,11	153)
Tetracene IV	31,12	27,80	1,12	154,90)
Pentacene V	31,36	27,80	1,13	153,45)
Pyrene VII	30,48	27,52	1,09	149,59)
Perylene VIII	30,64	28,28	1,08	154)
non-alternant:				
Acepleiadylene XIX	32,04	20,49	1,56	49)
Acenaphth[1,2-a]acenaphthylene XX	12,00	29,44	0,41	49)

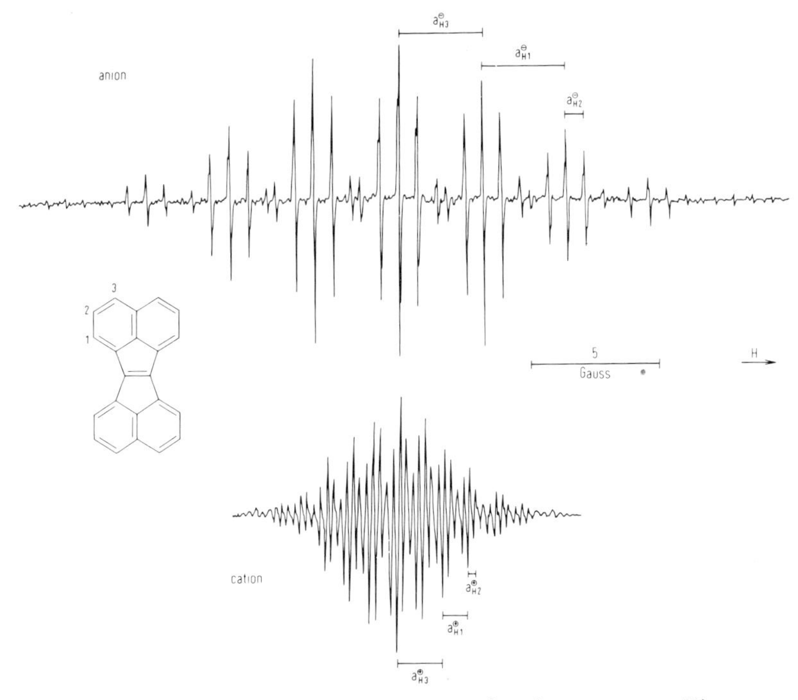

Figure 28: ESR spectra of the radical-ions of acenaphth [1,2-a]acenaphthylene[49].
Top: anion in 1,2-dimethoxyethane, at −70 °C, gegenion: $Na^{\oplus}$;
bottom: cation in conc. sulfuric acid, at +65 °C.

The great differences between the individual coupling constants $a^{\ominus}_{H\mu}$ and $a^{\oplus}_{H\mu}$ as well as those between the overall spectral ranges $\Sigma_\mu |a^{\ominus}_{H\mu}|$ and $\Sigma_\mu |a^{\oplus}_{H\mu}|$ are clearly visible in the spectra of the radical-anion and the radical-cation of acenaphth [1,2-a] acenaphthylene (XX)[49] reproduced in Figure 28.

2.2. Heterocyclic compounds

Aza-aromatic systems. The radical-ions of a large number of aza-aromatic compounds of the pyridine-type have been subjected to ESR spectroscopic investigations[31,33) 58,66,92,112-116,163,164,165b,166,167]. The coupling constants of ring protons and ^{14}N nuclei, together with the HMO spin populations $c^2_{a\mu}$ are given in Table 5 only for those radical-anions which do not contain adjacent nitrogen centres.

Table 5: Coupling constants of ring protons and ^{14}N nuclei in the radical-anions of aza-aromatic compounds, together with the corresponding HMO spin populations.

Radical anion of	μ	$a_{H\mu}$	$a_{N\mu}$	$c_{a\mu}^2$ *)	Ref.
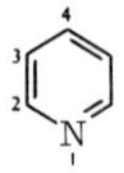 XXI Pyridine	1(N)		6,28	0,258	
	2	3,55		0,170	167)
	3	0,82		0,043	
	4	9,70		0,317	
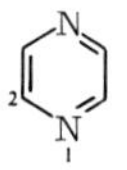 XXII Pyrazine	1(N)		7,18	0,260	
	2	2,64		0,120	112)
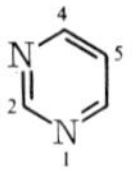 XXIII Pyrimidine	1(N)		3,26	0,148	
	2	0,72		0,000	167)
	4	9,78		0,351	
	5	1,31		0,000	
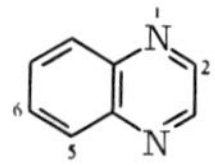 XXIV Quinoxaline	1(N)		5.64	0,215	
	2	3,32		0,139	
	5	2,32		0,075	32)
	6	1,00		0,048	
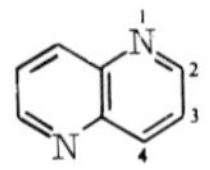 XXV 1,5-Naphthyridine	1(N)		3,36	0,130	
	2	2,95		0,118	166)
	3	1,69		0,060	
	4	5,78		0,182	

Radical anion of	μ	$a_{H\mu}$	$a_{N\mu}$	$c_{a\mu}^{2*)}$	Ref.
XXVI 1,8-Naphthyridine	1(N)		2,47	0,100	
	2	4,07		0,159	
	3	0,70		0,029	166)
	4	6,54		0,212	
XXVII 2,7-Naphthyridine	1	4,95		0,209	
	2(N)		3,40	0,134	166)
	3	0,43		0,001	
	4	4,42		0,127	
XXVIII 1,4,5,8-Tetra-azanaphthalene	1(N)		3,37	0,132	92)
	2	3,14		0,118	
XXIX Phenazine	1	1,93		0,054	
	2	1,61		0,049	32)
	9(N)		5,14	0,205	
XXX 1,4,5,8-Tetra-azaanthracene	1(N)		2,41.	0,102	
	2	2,73		0,094	32)
	9	3,96		0,107	

Radical anion of	μ	$a_{H\mu}$	$a_{N\mu}$	$c^2_{a\mu}$ *)	Ref.
XXXI 5,6,11,12-Tetra-azatetracene	1	0,84		0,026	166)
	2	1,40		0,042	
	5(N)		2,98	0,124	
XXXII 1,3,6,8-Tetra-azapyrene	1(N)		2,57	0,090	58)
	2	0,36		0	
	4	2,39		0,083	
XXXIII 4,4'-Dipyridyl	1(N)		3,64	0,130	32)
	2	0,43		0,058	
	3	2,35		0,065	
XXXIV 2,2'-Dipyridyl	1(N)		2,65	0,139	277)
	3	0,61		0,029	
	4	1,22		0,083	
	5	4,71		0,094	
	6	1,08		0,020	
XXXV 2,2'-Dipyrimidyl	1(N)		1,41	0,093	309)
	4	0,15		0,004	
	5	4,98		0,108	

*) Calculated with $a_N = \alpha + 0{,}89\beta$ and $\beta_{CN} = \beta$.

In the HMO model of these compounds, the bond integral was not modified for the C-N linkages relative to the C-C linkages ($\beta_{CN} = \beta$), while the value $\alpha_N = \alpha + 0.89\beta$ was used for the Coulomb integral of the nitrogen centres. This particular choice had been made, because it yields for the radical-anion of 1,4,5,8-tetraazanaphthalene (XXVIII) the HMO spin population $c_{a2}^2 = 0.118$ at the carbon centre 2. The ratio between this spin population and the corresponding value for the naphthalene (II) radical-anion (0.069) is the same as the ratio between the ring proton constants a_{H2} of XXVIII$^{\ominus}$ (3.14 gauss) and II$^{\ominus}$ (1.83 gauss) (cf. Table 2 and 5).

The regression of the coupling constants $a_{N\mu}$ of the ^{14}N nuclei on the calculated HMO spin populations $c_{a\mu}^2$ at the nitrogen centres μ (N) is shown in Figure 29.

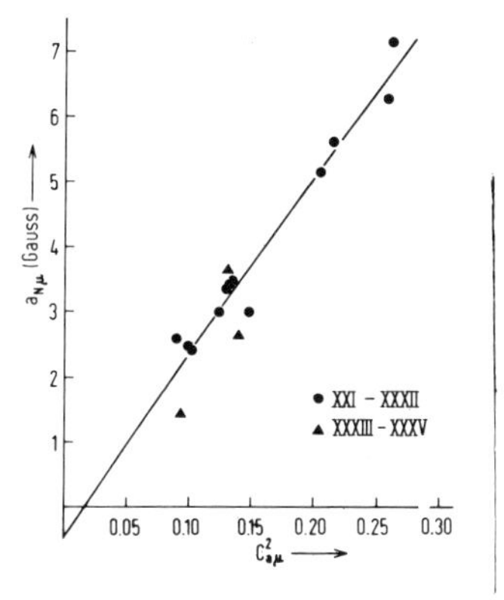

Figure 29: Regression of ^{14}N coupling constants $a_{N\mu}$ on the HMO spin populations $c_{a\mu}^2$ at nitrogen centres μ for the radical-anions of aza-aromatic compounds listed in Table 5.

The equation of this regression line is

$$a_{N\mu} = -0.45 + 27.6\, c_{a\mu}^2 , \tag{119}$$

and the standard deviation of individual measurements amounts to 0.42 gauss. This is similar to the value obtained in the case of the regression lines [eqs. (111) and (112)] considered in Section 2.1. Equation (119) corresponds to the approximation formula

$$a_{N\mu} = Q_N \rho_\mu^\pi , \tag{29}$$

in which no account is taken of the contributions of spin populations and ρ^{π}_{ν} and $\rho^{\pi}_{\nu'}$, at the adjacent centres ν and ν' (cf. Section 1.5). Such contributions are particularly important if the spin population ρ^{π}_{μ} at the nitrogen center μ is low, i.e. in the case of small coupling constants $a_{N\mu}$. This may be the reason for the significant deviation of the regression line from the origin of the coordinate system.

The empirical estimate of $|Q_N|$ from the slope of the regression line gives 27.6 ±2.0 gauss, this estimate being, of course, a function of the method used to calculate the spin populations. Other sets of HMO parameters, α_N and β_{CN} and/or refined MO methods lead to $|Q_N|$ values that differ from the one given above. Furthermore, the value of $|Q_N|$ depends on the selection of the experimental data $a_{N\mu}$. It is also expected to change when the approximation formula (29) is replaced by the more exact relationship

$$a_{N\mu} = Q_N \rho^{\pi}_{\mu} + Q_{CN}(\rho^{\pi}_{\nu} + \rho^{\pi}_{\nu'}) . \qquad (28)$$

In this case Q_N depends both on the magnitude and on the sign of the second parameter Q_{CN}. Whereas theory and experiment agree about the positive sign of Q_N (cf. Section 1.5), the sign of Q_{CN} involves considerable uncertainty. Thus, the proposed Q_{CN} values vary between +9 and −7 gauss, and the estimated values of Q_N vary accordingly between +19 and +31 gauss. (It is obvious that negative and positive Q_{CN} values lead to larger and smaller positive Q_N values, respectively.) One cannot yet with certainty decide which of the above parameters is the best to correlate the calculated spin populations with the measured coupling constants $a_{N\mu}$, but recent results[167] favour $Q_N \approx +27$ and $Q_{CN} \approx -2$ gauss.

In contrast to the case of the pyridine-type aza-aromatics, the nitrogen in cycl [3,2,2] azine[274] (XXXVI) contributes two electrons to the π-system. The coupling constants $a_{H\mu}$ and $a_{N\mu}$ of the ring protons and the ^{14}N nucleus in the radical-anion of XXXVI are given below:

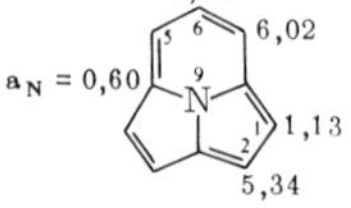

1,20
6,02
a_N = 0,60
1,13
5,34

XXXVI$^{\ominus}$

The nodal plane of the lowest antibonding HMO ψ_a in XXXVI traverses the nitrogen and the centre $\mu = 6$, so that the HMO values $c^2_{a\mu}$ vanish there. Refined methods (CI method and McLachlan's formula (44)) indicate negative spin populations at these two centres, the magnitudes of the calculated values ρ_9^π and ρ_6^π being in good agreement with the experimental coupling constants a_{N9} and a_{H6}.

Apart from a single exception to be discussed below, there are no known radical-cations of unprotonated aza-aromatic compounds, since these compounds exhibit basic properties, and only diamagnetic proton adducts are found in concentrated sulphuric acid used as an oxidant. An unprotonated paramagnetic species has been reported for 1,3,6,8-tetraazapyrene (XXXII). This species is formed in a small amount

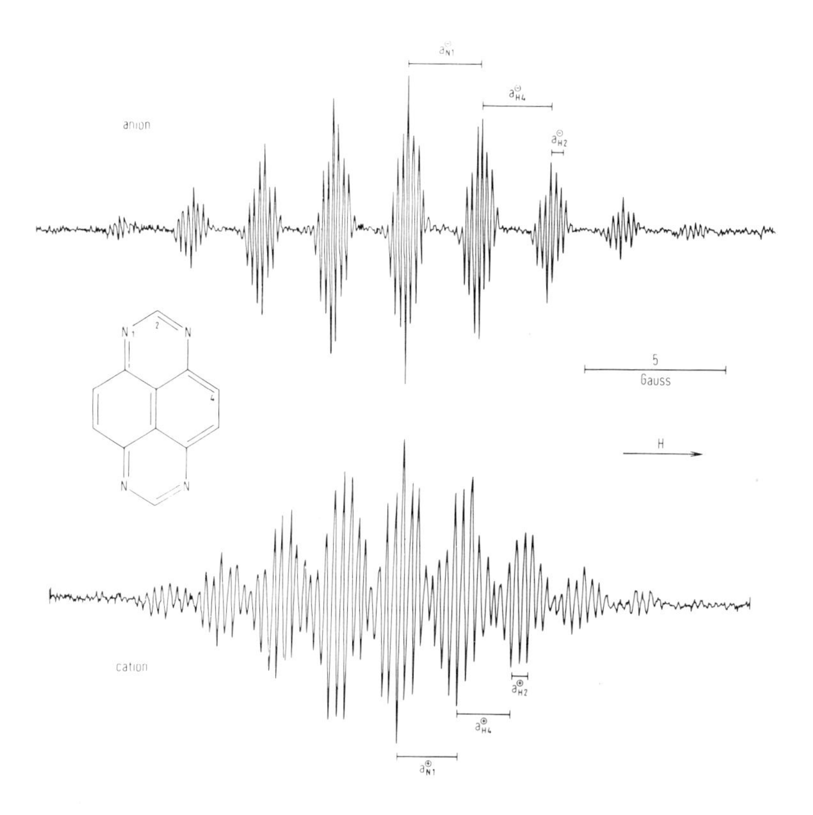

Figure 30: ESR spectra of the radical-ions of 1,3,6,8 -tetraazapyrene[58].
Top: anion in 1,2-dimethoxyethane, at +25°C; gegenion $K^\oplus$.
bottom: cation in conc.. sulfuric acid, at +80°C.

together with diamagnetic adducts when XXXII is dissolved in concentrated sulphuric acid, and may be regarded as the radical-cation of XXXII[58]. Its ESR experimental data do not differ much from those of the corresponding radical-anion: XXXII$^{\oplus}$: $a^{\oplus}_{N1} = 2.14$, $a^{\oplus}_{H2} = 0.54$, $a^{\oplus}_{H4} = 1.87$; XXXII$^{\ominus}$: $a^{\ominus}_{N1} = 2.57$, $a^{\ominus}_{H2} = 0.36$, $a^{\ominus}_{H4} = 2.39$ gauss. The similarity of the coupling constants of the two radical-ions is reflected in their ESR spectra which are reproduced in Figure 30.

HMO models show that this similarity is to be expected despite the perturbation of the alternant pyrene system, due to the replacement of the four carbon AO's ϕ_μ ($\mu = 1,3,6,8$) by nitrogen AO's. The centres $\mu = 2$ and $\mu = 7$ in XXXII$^{\ominus}$ and XXXII$^{\oplus}$ where the HMO quantities c^2_{j2} and c^2_{j7} are zero presumably exhibit negative spin populations, like the corresponding centres in the radical-ions of pyrene (cf. Appendix A 1.2).

Dihydro-derivates of aromatic diaza-compounds. As mentioned in Section 1.2, radical-cations may form when diprotonated diaza-compounds are reduced:

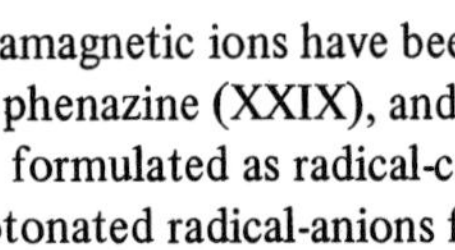

Such paramagnetic ions have been prepared from pyrazine (XXII), quinoxaline (XXIV), phenazine (XXIX), and 4,4'-bipyridyl (XXXIII)[60b,66,72,165a]. Though they are formulated as radical-cations of the corresponding dihydro-derivatives, they are diprotonated radical-anions from the viewpoint of their π-electron structure. Comparison between the relevant experimental data listed in Table 6 and those for XXII$^{\ominus}$, XXIV$^{\ominus}$, XXIX$^{\ominus}$ and XXXIII$^{\ominus}$ in Table 5 shows that similar $a_{N\mu}$ and $a_{H\mu}$ values appear in the two series, though the two added protons in the cations give rise to additional hyperfine structure. Protonation of the nitrogen does not reduce either the number of π-electrons or the number of centres, so that the π-systems of the cations differ from those of the corresponding anions mainly by the higher electronegativity of the nitrogen centres. Therefore, it seems justifiable to increase the parameter h_N of the Coulomb integral α_N from 0.89 to 1.2 in the calculation of

Table 6: Coupling constants of ring protons and ^{14}N nuclei in the radical-cations of the dihydro-derivatives of some aza-aromatic compounds, together with the corresponding HMO spin populations[72].

Radical-cation of the dihydro-derivative of	μ	$a_{H\mu}$	$a_{N\mu}$	$c^2_{a\mu}$ *)
Pyrazine	1(N)	7,94	7,40	0,232
XXII	2	3,13		0,134
Quinoxaline	1(N)	7,17	6,65	0,200
XXIV	2	3,99		0,161
	5	1,38		0,057
	6	0,78		0,046
Phenazine	1	0,66		0,044
XXIX	2	1,71		0,052
	9(N)	6,49	6,12	0,189
4,4'-Dipyridyl	1(N)	4,06	3,56	0,115
XXXIII	2	1,61		0,071
	3	1,45		0,059

*) Calculated with $\alpha_N = \alpha + 1{,}2\beta$ and $\beta_{CN} = \beta$.

the HMO spin populations (cf. Table 6). It must also be borne in mind that the unpaired electron occupies the HMO ψ_a, as in the radical-anions and not the HMO ψ_b as in the radical-cations discussed so far.

Thia-aromatic systems. Of the radical-ions of heteroaromatic compounds with a heteroatom other than nitrogen, only three 1,4-dithiin-type radical-cations will be mentioned here[56,57]. The coupling constants $a_{H\mu}$ of their ring protons are listed in Table 7, together with the HMO spin populations $c^2_{b\mu}$ at the corresponding carbon centres μ. These spin populations have been calculated with $\alpha_S = \alpha + \beta$ and $\beta_{CS} = 0.57\beta$, i.e. with the aid of a HMO model that treats the sulphur atom as a centre contributing one electron to the π-system.

Table 7: Coupling constants of ring protons and HMO spin populations in the radical-cations of three thia-aromatic compounds of the 1,4-dithiin type[316].

Radical-cation of	μ	$a_{H\mu}$	$c_{b\mu}^{2}$ *)
XXXVII 1,4-Dithiine	2	2,82	0,125
XXXVIII 1,4-Benzodithiine	2	3,32	0,158
	5	0,20	0,016
	6	1,05	0,032
XIXXX Thianthrene	1	0,14	0,015
	2	1,28	0,044

*) Calculated with $\alpha_S = \alpha + \beta$ and $\beta_{CS} = 0{,}57\beta$.

2.3. Quinones, aldehydes, and ketones

Quinones. The radical-anions of quinones, known as semiquione-anions appear as paramagnetic intermediates in the alkaline reduction of quinones to dihydroquinones, the latter being usually called hydroquinones[23].

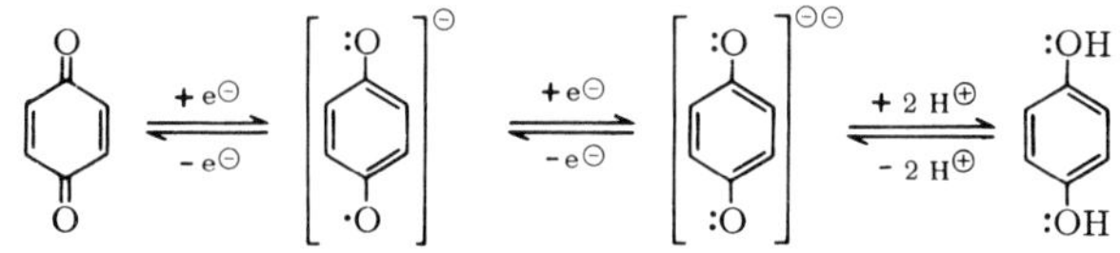

$+e^{\ominus}$ / $-e^{\ominus}$; $+e^{\ominus}$ / $-e^{\ominus}$; $+2\,H^{\oplus}$ / $-2\,H^{\oplus}$

Quinone — Semiquinone-anion — Dianion of Dithydroquinone — Dihydroquinone

The coupling constants $a_{H\mu}$ of the ring protons and the HMO spin populations $c^2_{a\mu}$ at the carbon centres μ are given in Table 8 for some semiquinone-anions [71,138,130,168]. Upon the suggestion of Vincow and Fraenkel[138], the values of $c^2_{a\mu}$ have been calculated with $\alpha_O = \alpha + 1.2\beta$ and $\beta_{CO} = 1.56\beta$.

The data listed in Table 8 show that this set of parameters applies to p-semiquinone-anions ($XL^{\ominus}$-$XLIV^{\ominus}$) but not to the o-semiquinone-anions ($XLV^{\ominus}$-$XLVII^{\ominus}$).

Table 8: Coupling constants of ring protons and HMO spin populations in some semiquinone-anions.

Semiquinone-anion of	μ	$a_{H\mu}$	$c^{2\,*)}_{a\mu}$	Ref.
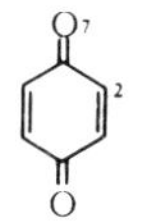 XL 1,4-Benzoquinone	2	2,37	0,107	[138]
XLI 1,4-Naphthoquinone	2	3,23	0,141	[138]
	5	0,51	0,025	
	6	0,66	0,032	

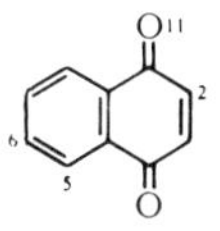

Semiquinone-anion of	μ	$a_{H\mu}$	$c^{2}_{a\mu}$ *)	Ref.
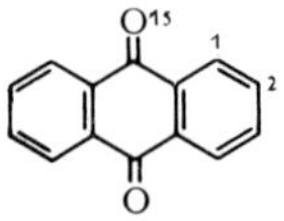 XLII 9,10-Anthraquinone	1	0,55	0,024	
	2	0,96	0,042	138)
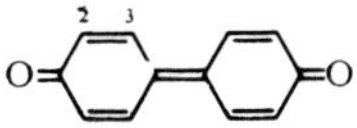 XLIII Diphenoquinone	2	2,29	0,82	130)
	3	0,53	0,34	
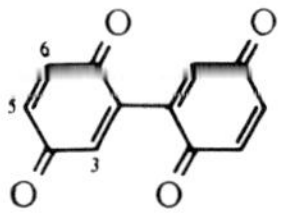 XLIV Di-1',4-benzoquinone	3	0,82	0,005	
	5	2,34	0,091	168b)
	6	1,13	0,062	
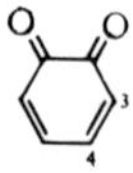 XLV 1,2-Benzoquinone	3	0,95	0,109	138)
	4	3,65	0,107	

Semiquinone-anion of	μ	$a_{H\mu}$	$c_{a\mu}^2$ *)	Ref.
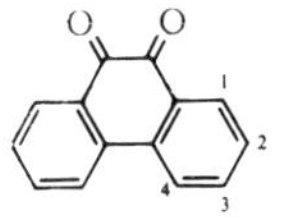 XLVI 9,10-Phenanthraquinone	1	1,35	0,032	
	2	0,20	0,021	
	3	1,67	0,044	71)
	4	0,42	0,010	
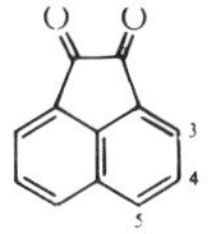 XLVII Acenaphthoquinone	3	1,17	0,025	
	4	0,27	0,000	71)
	5	1,27	0,027	

*) Calculated with $\alpha_O = \alpha + 1{,}2\beta$ and $\beta_{CO} = 1{,}56\beta$.

Aldehydes and ketones. The radical-anions of aldehydes and ketones are known as ketyl-anions. Like semiquinones, ketyls are paramagnetic intermediates arising in the reduction of the carbonyl group[21]:

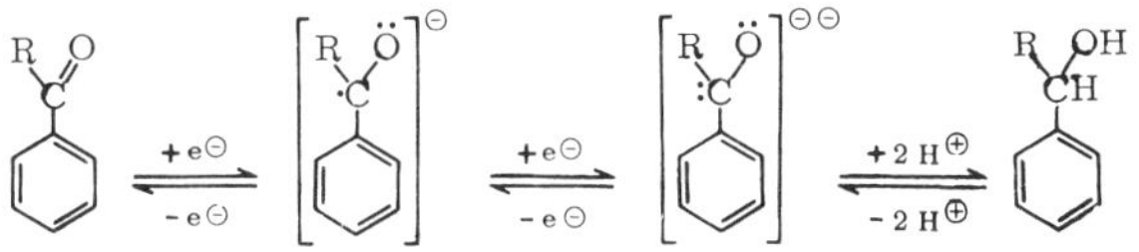

$+e^\ominus$ / $-e^\ominus$; $+e^\ominus$ / $-e^\ominus$; $+2\,H^\oplus$ / $-2\,H^\oplus$

Ketone or aldehyde (when R = H) — Ketyl-anion — "Dianion of the alcohol" — Alcohol

Table 9 contains the experimental coupling constants $a_{H\mu}$ for some ketyl-anions[37,69,71,169], together with the corresponding HMO spin populations $c_{a\mu}^2$ calculated with the same set of parameters as used for semiquinone-anions. The experimental and the calculated quantities are in moderately good agreement.

Table 9: Coupling constants of ring protons and HMO populations in some ketyl-anions.

Ketyl-anion of	μ	$a_{H\mu}$		$c^{2}_{a\mu}$ *)	Ref.
	2	4,69	4,04 average	0,123	
	6	3,39			
	3	1,31	1,03 average	0,018	169)
	5	0,75			
XLVIII	4	6,47		0,188	
Benzaldehyde	7	8,51		0,235	
	2	2,52		0,077	
	3	0,82		0,006	69)
	4	3,50		0,100	
XLIX Benzophenone					
	1	1,96		0,038	
	2	0,03		0,043	71)
	3	3,08		0,068	
	4	0,65		0,016	
L Fluorenone					
	1	3,06		0,088	
	2	0,03		0,034	71)
	3	2,12		0,056	
	9	0,10		0,004	
LI Phenanthrene-4,5-ketone					
	2	0,99		0,048	
	3	0,36		0,001	71)
	4	1,12		0,054	
LII Benzil**)					

*) Calculated with $\alpha_O = \alpha + 1{,}2\beta$ and $\beta_{CO} = 1{,}56\beta$

**) The ketyl-anion of benzil may be considered as a semidione substituted by two phenyl groups (cf. Chapter 2.6).

Dihydroquinones. The reduction of p-quinones in strongly acidic solution is accompanied by the formation of positively charged paramagnetic species known as semi-quinone-cations. These may be regarded either as diprotonated radical-anions of quinones, or as radical-cations of dihydroquinones.

⊕OH ... ⊕OH ⇌(+e⊖ / −e⊖) [•OH ... :OH]⊕ ⇌(+e⊖ / −e⊖) :OH ... :OH

Diprotonated quinone — Semiquinone-cation — Dihydroquinone

Table 10 shows the coupling constants $a_{H\mu}$ of the ring and hydroxyl protons as well as the HMO spin populations $c^2_{a\mu}$ for some semiquinone-cations[60].
As with the diaza-aromatic compounds discussed in Section 2.3, diprotonation neither modifies the number of π-electrons nor the number of the centres, but it raises the electronegativity of the heteroatoms. It also reduces the double bond character of the carbon-oxygen linkage. This is taken into account by calculating the HMO

Table 10: Coupling constants of protons and HMO spin populations in semiquinone-cations[60b].

Semiquinone-cation of dihydro-	μ	$a_{H\mu}$	$c^2_{a\mu}$ *)
1,4-Benzoquinone	7(O)	3,44	0,113
XL	2	2,36	0,090
1,4-Naphthoquinone	11(O)	2,42	0,079
XLI	2	3,20	0,106
	5	1,80	0,074
	6	0,86	0,042
9,10-Anthraquinone	15(O)	1,31	0,065
XLII	1	1,66	0,055
	2	1,07	0,042

*) Calculated with $\alpha_O = \alpha + 2{,}0\beta$ and $\beta_{CO} = 1{,}1\beta$.

spin populations $c^2_{a\mu}$ with a new set of parameters[60b] : $\alpha_O = \alpha + 2.0\beta$ and $\beta_{CO} = 1.1\beta$. It is again important to recall here that in the semiquinone-cations, as in the corresponding unprotonated anions, the unpaired electron occupies the lowest vacant HMO ψ_a of quinone.

2.4. Substituted compounds

The various substituents of aromatic radical-ions differ fundamentally in their influence on the distribution of the spin population in the original unsubstituted π-electron system. One thus distinguishes between two types of substituents:

a) substituents strongly conjugated with the aromatic system and sharing appreciably in the spin population of the radical-ion;
b) substituents weakly conjugated with the aromatic system and modifying only slightly the original spin population *).

Group (a) comprises notably the nitro and the amino substituents, while alkyl substituents are typical representatives of group (b). Most other substituents (e.g. the cyano substituent and the halogens) are intermediate between the two groups.

Nitro-substituted systems. The radical-anions of nitrobenzene and of the polynitro-derivates of benzene have been investigated by numerous authors[24b,38,39,64,171-183]. The coupling constants of the ring protons and the ^{14}N nuclei are given below for some of these radical-anions generated electrolytically in polar solvents[64,173,181]. The ESR spectrum of nitrobenzene radical-anion $LIII^{\ominus}$ has been shown in Figure 24 (Appendix A 1.3).

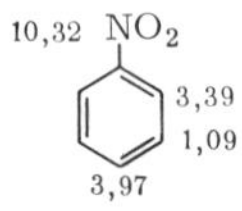

10,32 NO_2 3,39 1,09 3,97
$LIII^{\ominus}$

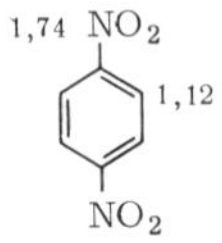

1,74 NO_2 1,12 NO_2
$LIV^{\ominus}$

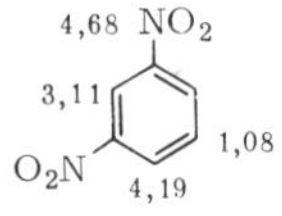

4,68 NO_2 3,11 O_2N 1,08 4,19
$LV^{\ominus}$

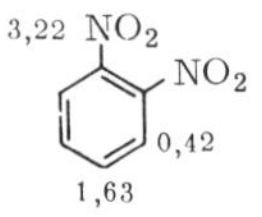

3,22 NO_2 NO_2 0,42 1,63
$LVI^{\ominus}$

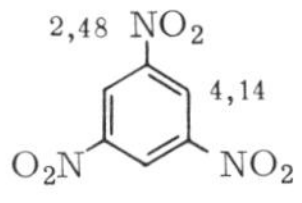

2,48 NO_2 4,14 O_2N NO_2
$LVII^{\ominus}$

*) Except for cases in which the unsubstituted radical-ion is in a degenerate ground state (cf. Section 2.5).

The association between alkali metal cations and the radical-anions $LIV^{\ominus}$, $LV^{\ominus}$ and $LVII^{\ominus}$ in etheral solvents gives rise to unusual ^{14}N-hyperfine structures[172c)] (cf. Appendix A 2.2): The experimental coupling constants $a_N^{NO_2}$ appear to be much larger than for the non-associated radical-anions, and also to be due to a single ^{14}N nucleus. It has later been found that the association leads to reduction of symmetry and complete redistribution of the spin population[172b,315)]. The ion-pairs of the radical-anion $LV^{\ominus}$ of m-dinitrobenzene with $\overset{\oplus}{Na}$ have been particularly well studied. The coupling constants of ^{14}N nuclei and ring protons in these ion-pairs are compared below with the corresponding values of the non-associated species.

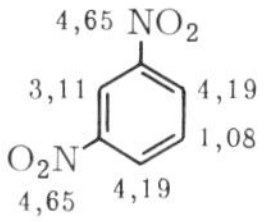

4,65 NO_2; 3,11; 4,19; 1,08; O_2N 4,65; 4,19

$LV^{\ominus}$

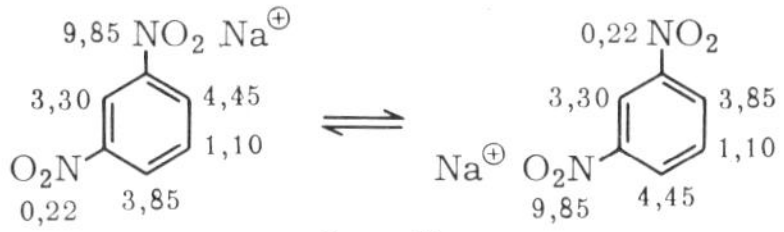

9,85 NO_2 $Na^{\oplus}$; 3,30; 4,45; 1,10; O_2N 0,22; 3,85 ⇌ 0,22 NO_2; 3,30; 3,85; 1,10; $Na^{\oplus}$ O_2N 9,85; 4,45

$LV^{\ominus}$-$Na^{\oplus}$

non-associated radical-anion	ion-pairs
(in acetonitrile, at +25 °C; gegenion: tetra-n-propylammonium$^{\oplus}$)[173)]	(in 1,2-dimethoxyethane, at +25 °C; gegenion: $Na^{\oplus}$)[315)]

The interconversion of the equivalent ion-pairs is brought about by the oscillation of the cation between two equivalent sites in close proximity to the nitro groups. This oscillation gives rise to anomalous line-widths when the time spent by the cation at one site is of the order of 10^{-7} sec[315)] (cf. Appendix A.2.3).

The results obtained with electrolytically produced radical-anions of nitrobenzene (LIII) and of its substituted derivatives also demonstrate the great sensitivity of the ^{14}N coupling constants to the solvent and the gegenion[174,175)]. It has thus been found

that this coupling constant increases by about 50% when the water content of the solvent (N,N-dimethylformamide) is raised from 0 to 80-90%, while the ring proton coupling constants exhibit a much smaller change[174a,b)]. Replacement of tetra-alkylammonium-cations in the supporting electrolyte by metal ions also leads to a change in the same direction[174c)].

Of the few known radical-anions of nitro-substituted naphthalenes[94,184)] the two peri-derivatives $LVIII^{\ominus}$ and $LIX^{\ominus}$ are of particular interest, because the NO_2 groups in these are not coplanar with the aromatic ring.

3,03
O_2N NO_2
3,63
0,95
3,73

LVIII$^{\ominus}$

0,20
O_2N NO_2
1,48
O_2N NO_2

LIX$^{\ominus}$

In connection with LIX$^{\ominus}$, one should note the small value of the ^{14}N coupling constant, and also the fact that no unusual hyperfine structure is exhibited by ethereal solutions in the presence of alkali metal cations as gegenions. In this respect LIX$^{\ominus}$ differs from LVIII$^{\ominus}$ and other radical-anions of polynitro-compounds[94].

The coupling constant $a_N^{NO_2}$ of the ^{14}N nucleus in a nitro substituent can be related to the π-spin populations by an expression similar to eq. (28), i.e.

$$a_N^{NO_2} = Q_N \rho_N^\pi + Q_{CN} \rho_\mu^\pi + 2\, Q_{ON} \rho_O^\pi \,. \tag{120}$$

Accordingly, the value of $a_N^{NO_2}$ depends on the spin population ρ_N^π at the nitrogen and the spin populations $\rho_{\tilde{\mu}}^\pi$ and ρ_O^π at the substituted carbon centre $\tilde{\mu}$ and the two oxygen atoms, respectively.

Rieger and Fraenkel[176] have calculated the spin populations in the radical-anions of some nitro-derivatives of benzene, using the McLachlan procedure (44) and starting with the following set of HMO parameters:

$$\alpha_N = \alpha + 2.2\,\beta;\ \alpha_O = \alpha + 1.4\,\beta;\ \beta_{CN} = 1.2\,\beta\text{: and }\ \beta_{NO} = 1.67\,\beta\,.$$

These parameters lead to satisfactory agreement between the spin populations ρ_μ^π at the proton-carrying centres μ and the coupling constants $a_{H\mu}$ of the ring protons.

Substitution of the calculated values ρ_N^π, $\rho_{\tilde{\mu}}^\pi$ and ρ_O^π into eq. (120) yields those empirical estimates of Q_N, Q_{CN} and Q_{ON} which give the best possible correlation with the observed coupling constants $a_N^{NO_2}$. It has been found that in eq. (120), too, the parameter Q_{CN} is negligibly small, so that the simplified approximation formula

$$a_N^{NO_2} \approx Q_N \rho_N^\pi + 2\, Q_{ON} \rho_O^\pi \tag{121}$$

may be used, where $|Q_N|$ and $|Q_{ON}|$ have the values of 99 ± 10 gauss and 36 ± 6 gauss, respectively[176], and differ in sign[175b]. Although these parameters give a good agreement also for the radical-anions of aliphatic nitro-compounds (cf. Section 1.2), completely different Q_N- and Q_{NO} - values have recently been proposed [184,185].

According to the theoretical and experimental evidence available (cf. Appendix A 1.3), the set $Q_N > 0$, $Q_{ON} < 0$ is certainly more likely than the reverse[176,184].

Amino-substituted systems. Whereas the nitro group is a powerful electron-acceptor and raises the ionization potential of aromatic compounds, the amino group is an electron-donor and is one of the substituents that decrease this potential most efficiently. It is therefore to be expected that, unlike the nitro-compounds (of which only the radical-anions can be prepared), the amines will yield predominantly radical-cations.

The stable radical-cation of N,N,N',N'-tetramethyl-p-phenylenediamine (LXI), known as Wurster's Blue, was one of the first radical-ions to be investigated by ESR spectroscopy[24a,b,84,186]. The radical-cation of p-phenylenediamine itself (LX) is less stable than LXI$^{\oplus}$. The coupling constants of the protons and the ^{14}N nuclei in LX$^{\oplus}$ and LXI$^{\oplus}$ are shown below, together with those of the radical-cations LXII$^{\oplus}$ and LXIII$^{\oplus}$ obtained from benzidine and N,N,N',N'-tetramethylbenzidine[65,77,186b], respectively.

LX$^{\oplus}$: 2,13 (ring H); 5,88 (NH); 5,29 (N)

LXII$^{\oplus}$: 1,08; 1,62 (ring H); 3,97 (NH); 3,60 (N)

LXI$^{\oplus}$: 1,98 (ring H); 6,74 (CH_3); 7,02 (N)

LXIII$^{\oplus}$: 0,73; 1,66 (ring H); 4,70 (CH_3); 4,88 (N)

The coupling constants $a_N^{NH_2}$ and $a_N^{N(CH_3)_2}$ of the ^{14}N nuclei in the amino groups can be expressed as

$$a_N^{NH_3} \text{ or } a_N^{N(CH_3)_2} = Q_N \rho_N^{\pi} + Q_{CN} \rho_{\mu}^{\pi}, \qquad (122)$$

where ρ_N^{π} and ρ_{μ}^{π} denote the spin populations at the nitrogen and the substituted carbon centre, respectively[77]. By neglecting the contribution of ρ_{μ}^{π} ($|Q_{CN}| \ll |Q_N|$) one obtains the approximation formula:

$$a_N^{N_2} \text{ or } a_N^{N(CH_3)_2} \approx Q_N \rho_N^{\pi}. \qquad (123)$$

Eqs. (122) and (123) are analogous to eqs. (28) and (29) derived for the coupling constants $a_{N\mu}$ of the ^{14}N nuclei in the radical-anions of aza-aromatic compounds (cf. Section 1.5 and 2.2). There is also no considerable difference between the values of Q_N obtained in the two cases (+20 to +30 gauss). These values again depend on the method used to calculate the spin populations, and on whether or not the contributions of the spin populations at the adjacent carbon centres are neglected[77) [eq. (122) or eq. (123)].

Alkyl-substituted systems. The radical-ions whose ESR-data and π-spin populations have so far been discussed have no alkyl groups attached to the aromatic ring. However, alkyl- (particularly methyl-) derivatives of many of the radical-ions mentioned before are now known. These include radical-ions of alkyl-derivatives of benzene (I)[30,80,187-190], naphthalene (II)[52,53,82,123,130-132,191], anthracene (III)[50,51] biphenylene (X)[192], cyclooctatetraene (XII)[193], biphenyl (XIII)[194], azulene (XVI)[67] heterocycles XXI, XXII, XXIV and XXXIX[56b,72,112,114,167], 1,4-benzoquinone (XL)[25b,168a,196,197,297], benzophenone (XLIX)[198], nitrobenzene (LIII)[38,39,176,177,199] and polynitrobenzenes LIV, LV and LVII[172c,179a,c,180].

It is impossible to discuss here the extensive experimental material, and only representative examples will be considered.

The radical-anions of alkyl-substituted aromatic hydrocarbons have been thoroughly investigated. Amongst these, the radical-anions of alkyl-substituted π-electron perimeters (benzene, cyclooctatetraene) constitute a special case and their discussion will be postponed to Section 2.5.

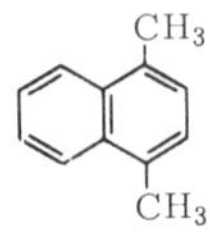

LXIV (1,4) 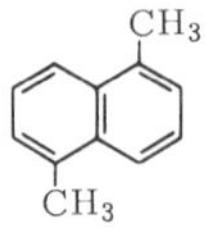 LXIV (1,5) 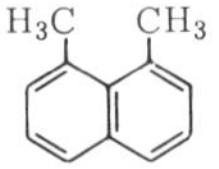 LXIV (1,8)

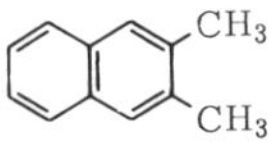

LXIV (2,3) 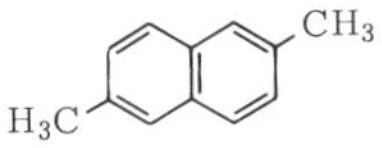 LXIV (2,6) 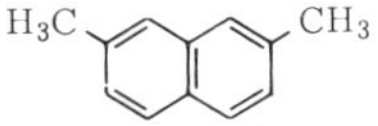 LXIV (2,7)

At this point, the radical-anions of the six symmetrically substituted dimethylnaphthalenes LXIV ($\tilde{\mu}, \tilde{\mu}'$) will be dealt with. Their experimental data are suitable for checking the spin populations calculated with the aid of a modified HMO model.

Table 11: Coupling constants of ring and methyl protons in the radical-anions of the six symmetrically substituted dimethylnaphthalenes[82] LXIV (μ, $\tilde{\mu}$), together with the corresponding HMO spin populations.

$\tilde{\mu},\tilde{\mu}$	μ	$a_{H\mu}$ *)	$c^2_{a\mu}$ **)	$\tilde{\mu},\tilde{\mu}'$	μ	$a_{H\mu}$ *)	$c^2_{a\mu}$ **)
	2	1,63	0,047		1	4,67	0,153
	5	5,17	0,220		5	4,93	0,188
1,4	6	1,79	0,075	2,3	6	1,76	0,068
	1($\tilde{\mu}$)	3,26(CH_3)	0,138		2($\tilde{\mu}$)	1,69(CH_3)	0,081
	H_3		0,0158		H_3		0,0092
	2	1,13	0,052		1	4,65	0,163
	3	2,46	0,070		3	2,68	0,107
1,5	4	4,50	0,174	2,6	4	4,79	0,178
	1($\tilde{\mu}$)	4,41(CH_3)	0,183		2($\tilde{\mu}$)	1,22(CH_3)	0,044
	H_3		0,0210		H_3		0,0050
	2	1,70	0,038		1	4,32	0,148
	3	1,70	0,086		3	1,76	0,084
1,8	4	4,73	0,165	2,7	4	5,12	0,196
	1($\tilde{\mu}$)	4,61(CH_3)	0,189		2($\tilde{\mu}$)	2,16(CH_3)	0,064
	H_3		0,0217		H_3		0,0072

*) (CH_3): coupling constant $a_H^{CH_3}$ of a methyl proton.
**) Calculated with $\alpha_{\tilde{\mu}} = \alpha - 0{,}3\beta$; $\alpha_M = \alpha$; $\alpha_{H_3} = \alpha - 0{,}5\beta$; $\beta_{\tilde{\mu}M} = \beta$; and $\beta_{MH_3} = 3\beta$ (see text).

Table 11 contains the coupling constants $a_{H\mu}$ and $a_H^{CH_3}$ of the ring and methyl protons[82], together with the corresponding HMO spin populations $c^2_{a\mu}$ for the radical-anions $LXIV^{\ominus}(\tilde{\mu}, \tilde{\mu}')$. A modified HMO model described in Section 1.6 was used in which the hyperconjugation of the methyl groups with the naphthalene ring was accounted for by the following parameters:

$$C_{\tilde{\mu}}-C_M\equiv H_3 \qquad \alpha_M = \alpha \qquad \beta_{\tilde{\mu}M} = \beta$$
$$\alpha_{H_3} = \alpha - 0{,}5\beta \qquad \beta_{MH_3} = 3\beta$$

To allow for the inductive effect of the methyl groups, the Coulomb integral of the substituted centres $\tilde{\mu}$ was set equal to

$$\alpha_{\tilde{\mu}} = \alpha - 0.3\,\beta.$$

Figure 31 shows the regression of the coupling constants $a_H^{CH_3}$ of the methyl protons on the calculated spin populations $c_{aH_3}^2$ $(\approx \rho_{H_3}^{ls})$.

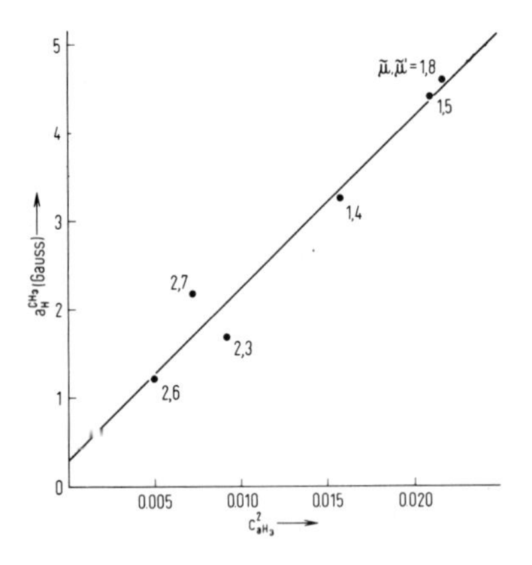

Figure 31: Regression of the methyl proton coupling constants $a_H^{CH_3}$ on the HMO spin populations $c_{aH_3}^2$ at the pseudo-centres H_3 for the radical-anions of the six symmetrically substituted dimethylnaphthalenes listed in Table 11.

The equation of the regression line is

$$a_H^{CH_3} = 0.30 + 194\, c_{aH_3}^2 \tag{124}$$

and the standard deviation of individual measurements amounts to 0.32 gauss. The slope (194 gauss) does not differ too much from the value expected on the basis of the relationship

$$a_H^{CH_3} = \frac{1}{3} Q_H \cdot \rho_{H_3}^{1s}, \tag{31}$$

where Q_H is taken to be +510 gauss (cf. Section 1.5).

A similarly good correlation as between $a_H^{CH_3}$ and $c_{aH_3}^2$ is obtained when the coupling constants $a_H^{CH_3}$ are plotted against the HMO spin populations $c_{a\tilde{\mu}}^2 \approx \rho_{\tilde{\mu}}^{\pi}$ at the substituted centres $\tilde{\mu}$, since the values $c_{aH_3}^2$ are closely proportional to $c_{a\tilde{\mu}}^2$ (cf. Table 11). This plot corresponds to the approximation formula

$$a_H^{CH_3} \approx Q_{CCH_3} \rho_{\tilde{\mu}}^{\pi} \tag{32}$$

for freely rotating methyl groups and not too different π-charges $q_{\tilde{\mu}}^{\pi}$ at the substituted centres $\tilde{\mu}$ (cf. Section 1.5). The empirical estimate for $|Q_{CCH_3}|$, which is derived from the regression of $a_H^{CH_3}$ on $c_{a\tilde{\mu}}^2$, amounts to 22.4 ± 2.3 gauss. This value is considerably smaller than that found for the coupling constant of methyl protons in the neutral ethyl radical (26.9 gauss)[119] and frequently used as an estimate of $|Q_{CCH_3}|$. As mentioned in Section 1.5, Q_{CCH_3} is not a constant, but a function of the charge $\rho_{\tilde{\mu}}^{\pi}$ at the substituted centre $\tilde{\mu}$. The effect of $q_{\tilde{\mu}}^{\pi}$ on the magnitude of the coupling constants $a_H^{CH_3}$ or $a_H^{CH_2}$ of β-alkyl protons can be best demonstrated with the radical-anions and cations derived from the same alkyl-substituted alternant hydrocarbons. Despite the perturbations caused by the introduction of an alkyl group, the corresponding radical-ions generally exhibit similar π-spin populations ($\rho_{\mu}^{\pi\ominus} \approx \rho_{\mu}^{\pi\oplus}$), and accordingly, the coupling constants $a_{H\mu}^{\ominus}$ and $a_{H\mu}^{\oplus}$ of their ring protons differ only slightly. However, since the π-charges at the substituted centres are very different for the anions and the cations ($q_{\tilde{\mu}}^{\pi\ominus}>1$; $q_{\tilde{\mu}}^{\pi\oplus}<1$;) the coupling constants $a_H^{CH_3\oplus}$ or $a_H^{CH_2\oplus}$ are much greater than $a_H^{CH_3\ominus}$ or $a_H^{CH_2\ominus}$. This can be seen from Table 12, which contains the experimental values for the radical-ions of pyracene (LXV)[52,131] and 9,10-dimethylanthracene (LXVI)[51], i.e. alkyl-derivatives of the alternant systems naphthalene (II) and anthracene (III). Table 12 also includes the relevant HMO spin populations $c_{a\mu}^2 = c_{b\mu}^2$ in the unsubstituted radical-ions of II and III. Furthermore, listed in the same table, for comparison's sake, are the analogous theoretical and experimental values of the radical-ions of acepleiadiene (LXVII)[55,305] and 3,5,8,10-tetramethylcyclopenta[ef]heptalene (LXVIII)[55,306], two alkyl-substituted, non-alternant hydrocarbons. For the radical-anion and the radical-cation of such a system the spin populations are different ($c_{a\mu}^2 \neq c_{b\mu}^2$), and consequently the coupling constants of both ring and alkyl protons differ greatly for the corresponding radical-ions of LXVII and LXVIII. The influence of $q_{\tilde{\mu}}^{\pi}$ on $a_H^{CH_3}$ was recently examined in a more quantitative manner by Hulme and Symons[80b)] on the basis of experimental data. It was mentioned in Section 1.2 that these authors succeeded in preparing the radical-cations from the alkyl-derivatives LXIX and LXX of benzene[80].

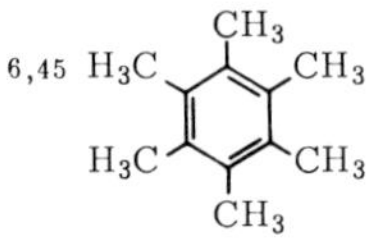

6,45 H_3C, CH_3 (hexamethylbenzene)

LXIX$^{\oplus}$

11,0 H_3C, <0,5 (ring H)

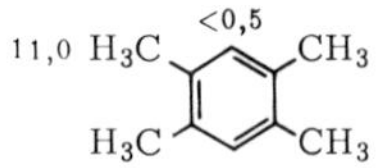

LXX$^{\oplus}$

Table 12: Coupling constants of the ring and alkyl protons in some alkyl-substituted hydrocarbon radical-ions[51,52,55,131], together with the HMO spin populations of the corresponding unsubstituted species.

Radical ion of	μ	$a_{H\mu}^{\ominus}$ *)	$a_{H\mu}^{\oplus}$ *)	$c_{a\mu}^2$ **)	$c_{b\mu}^2$ **)
Pyracene LXV	1	6,58(CH_2)	12,80(CH_2)	0,181	
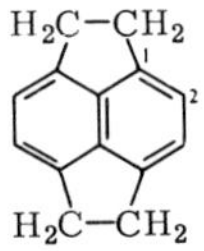	2	1,58	2,00	0,069	
9,10-Dimethyl-anthracene LXVI	1	2,90	2,54	0,119	
	2	1,52	1,19	0,031	
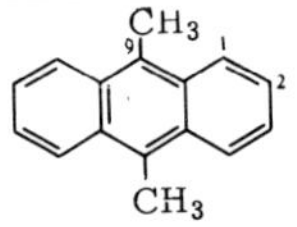	9	3,88(CH_3)	8,00(CH_3)	0,259	
Acepleiadiene LXVII	1	6,33	2,44	0,238	0,104
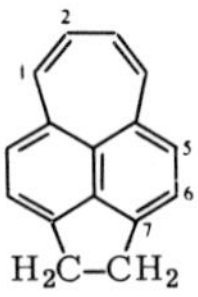	2	2,56	2,10	0,112	0,068
	5	0,71	3,50	0,032	0,135
	6	0,20	0,17	0,001	0,009
	7	3,05(CH_2)	10,06(CH_2)	0,037	0,152

Radical ion of	μ	$a_{H\mu}^{\ominus}$ *)	$a_{H\mu}^{\oplus}$ *)	$c_{a\mu}^2$ **)	$c_{b\mu}^2$ **)
3,5,8,10-Tetramethyl-cyclopenta-[ef] heptalene	1	0,32	2,07	0,005	0,068
	3	5,44(CH_3)	0,35(CH_3)	0,158	0,033
	4	1,28	4,99	0,001	0,135
	5	5,12(CH_3)	1,22(CH_3)	0,149	0,009
	6	0,96	6,07	0,027	0,152

H_3C 1 3 CH_3 4 H_3C 6 5 CH_3

LXVIII

*) (CH_3) and (CH_2): coupling constant of a methyl or methylene proton respectively.
**) Calculated for unsubstituted systems: $\alpha_{\tilde{\mu}} = \alpha$.

Halogen- and cyano-substituted systems. It is obvious that reduction by alkali metals can not be used as a method of preparation of halogen-substituted radical-anions. A great number of such radical-anions have been prepared electrolytically from the chloro-derivatives of 1,4-benzoquinone (XL) [196a,297,298] and from the chloro- and fluoro-derivatives of nitrobenzene (LIII)[38,39,174b,175a,176,177,199,202]. However, the radical-anions of halogen-substituted benzenes are unknown, because the low electron affinity of the parent neutral compounds precludes the electrolytic generation of the radical-anions.

Since the chlorine isotopes ^{35}Cl and ^{37}Cl possess only small magnetic moments (cf. Table 1, Section 1.1), the hyperfine splitting resulting from these nuclei generally is not resolved. No reliable direct comparison can therefore be made between the experimental data and the spin populations calculated for chlorine substituents. On the other hand, the ^{19}F nucleus ($I = \frac{1}{2}$) gives rise to appreciable splitting, which may be expressed by

$$a_F = Q_F \rho_F^\pi + Q_{CF} \rho_{\tilde{\mu}}^\pi, \tag{125}$$

analogously to the coupling constants of ^{13}C and ^{14}N nuclei [cf. eqs. (24) and (25)]. ρ_F^π and $\rho_{\tilde{\mu}}^\pi$ in eq. (125) refer to the spin populations at the fluorine atom and at the substituted carbon centre $\tilde{\mu}$.

Investigations of the radical-anions of three isomeric monofluoro-substituted nitrobenzenes (LXXI, LXXII, and LXXIII) have shown that the introduction of fluorine into the benzene ring causes no substantial redistribution of the spin population either in the ring or at the nitro group[202].

14,15 NO_2; F 6,53; 3,43; 1,17; 1,00; 3,60 — LXXI$^{\ominus}$

13,30 NO_2; 3,48; 3,48; 1,16; F 3,08; 3,67 — LXXII$^{\ominus}$

14,43 NO_2; 3,46; 1,13; 8,04 F — LXXIII$^{\ominus}$

The coupling constants a_F of the ^{19}F nuclei are about twice as large as the coupling constants $a_{H\mu}$ of the ring protons in the corresponding positions of the nitrobenzene radical-anion. This leads to the approximation formula

$$a_F \approx Q'_{CF}\, \rho^{\pi}_{\mu}\,, \tag{126}$$

which is compatible with the more exact relationship (125) if the ratio $\rho^{\pi}_{\mu} / \rho^{\pi}_{F}$ of the spin populations in LXXI$^{\ominus}$, LXXII$^{\ominus}$, and LXXIII$^{\ominus}$ remains roughly constant. The estimated value of $|Q'_{CF}|$ is 50 ± 10 gauss[202,320]. In contrast to Q_{CH}, the sign of Q_{CF} should be positive.

Radical-anions containing cyano-substituents have been produced from the cyano-derivatives of benzene (I), biphenyl (XIII), pyridine (XXI), nitrobenzene (LIII), and toluene (LXXVIII)[68,70,176,199,203,204]. The experimental data for the radical-anions of benzonitrile (LXXIV) and the three isomeric dicyanobenzenes (LXXV, LXXVI, and LXXVII) are given below[70]:

N 2,15; 3,63; 0,30; 8,42 — LXXIV$^{\ominus}$

N 1,81; 1,59 — LXXV$^{\ominus}$

N 1,02; 1,44; 8,29; < 0,08 — LXXVI$^{\ominus}$

N 1,75; 4,13; 0,42 — LXXVII$^{\ominus}$

The coupling constant a_N^{CN} of the ^{14}N nucleus in a cyano group can be expressed by

$$a_N^{CN} = Q_N\, \rho^{\pi}_{N} + Q_{CN}\, \rho^{\pi}_{C}\,, \tag{127}$$

where ρ^{π}_{N} and ρ^{π}_{C} denote the spin populations at the nitrogen and the carbon atom of the cyano group. The calculation of the HMO and McLachlan spin populations with the parameters

$$\alpha_N = \alpha + 1.0\beta;\ \beta_{CN} = 2.0\beta;\ \text{and}\ \beta_{\bar{\mu}C} = 0.9\beta\ (\bar{\mu} = \text{substituted centre})$$

leads to the values $|Q_N| = 23 \pm 2$ gauss and $|Q_{CN}|=7\pm 2$ gauss. These two values should differ in sign[68], Q_N being probably positive and Q_{CN} thus negative.

2.5. π - Electron perimeters

Fully conjugated rings of sp^2-hybridized carbon atoms are conveniently treated as π-electron perimeters. These systems are of particular theoretical interest, because, in the case of the highest symmetry of the ring (regular polygon), the molecular orbitals are determined completely by symmetry,and not by the type of approximation method used. This is so with two of the perimeters to be discussed in this Section: benzene and planar cyclooctatetraene. Therefore, the HMO's and the more accurate MO's are identical in these two systems. The HMO's of the alternant perimeters, in which the number of the centres is even and amounts to 2m, exhibit the following properties[28]:

1. Except for the lowest bonding orbital (ψ_1) and the highest antibonding orbital (ψ_{2m}), all the HMO's are doubly degenerate.

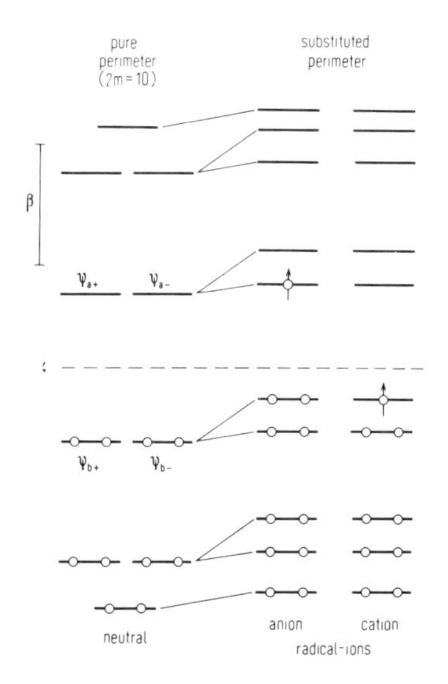

Figure 32: Energy diagram of a pure (4r + 2)-perimeter and an alkyl-substituted one (ten-membered ring). Occupancy of orbitals in the pure neutral perimeter and in the radical-ions of the substituted perimeter.

2. Two degenerate HMO's with energy E_i can be classified either as symmetric (ψ_{i+}) or as antisymmetric (ψ_{i-}) relative to a plane that traverses two opposite centres μ in the perimeter and is perpendicular to the nodal plane of the π-electron system.

3. The symmetrical HMO's ψ_{i+} and $\psi_{\hat{i}+}$ with energy $E_i = \alpha + x_i\beta$ and $E_{\hat{i}} = \alpha - x_i\beta$, respectively,are connected by the pairing properties (cf. Section 1.6).Such properties also connect the corresponding antisymmetric HMO's ψ_{i-} and $\psi_{\hat{i}-}$.

4. When the number 2m of the centres is a multiple of 4 (2m = 4r, where r = 1, 2, 3 . . .), then the perimeter possesses two degenerate non-bonding HMO's (ψ_{n+} and ψ_{n-}). Perimeters containing 2m = 4r + 2 centres lack such HMO's. According to this model, a neutral perimeter should be stable only in this second case (Hückel's rule[28,210]).

(4r + 2)-Perimeters. As can be seen from Figure 32, both the lowest vacant antibonding HMO's (ψ_{a+} and ψ_{a-}) and the highest occupied bonding HMO's (ψ_{b+} and ψ_{b-}) of a (4r +2)-perimeter are degenerate. Therefore, the probability that an unpaired electron in the radical-anion of such a perimeter is delocalized according to ψ^2_{a+} is exactly equal to the probability of its being delocalized according to ψ^2_{a-}. The same applies to the unpaired electron in the corresponding radical-cation with respect to ψ^2_{b+} and ψ^2_{b-}. The ground state of the radical-ions derived from the perimeter is thus doubly degenerate*).

When the contributions of the degenerate HMO's (ψ^2_{a+} and ψ^2_{a-} in an anion, and ψ^2_{b+} and ψ^2_{b-} in a cation) are given the same weight, then equal spin populations $\rho^\pi_\mu = 1/2m$ should prevail at all centres μ in the radical-ion of a perimeter. This is also expected on pure symmetry grounds for the radical-ions possessing the full symmetry D_{2mh}, as has been confirmed in the case of benzene (2m = 6) by the ESR spectrum of its radical-anion[30].

The introduction, into the perimeter, of a substituent exerting a weak and essentially inductive effect on the π-electron system causes only a small perturbation in the HMO energies. However, in most cases, such a perturbation affects differently the energies of the symmetric HMO's (ψ_{a+} and ψ_{b+}) and those of the antisymmetric ones (ψ_{a-} and ψ_{b-}) so that the degeneracy will be lifted (cf. Figure 32).

*) The twofold degeneracy in the ground state of radical-ions is often indicated by the width of ESR hyperfine lines[30,36,46,205] and their behaviour on saturation by high intensities of microwave irradiation (cf. Section 1.3). This is explained as follows: Non-linear molecules in a degenerate ground state are subject to Jahn–Teller distortions[206], which reduce the symmetry of the molecule and thus lift the degeneracy. The resulting variations in the coupling constants broaden the hyperfine lines[207]. Moreover, the existence of two degenerate or nearly degenerate states should constitute an extra means of spin-lattice relaxation[208], so that these hyperfine lines are expected to show saturation phenomena at much higher microwave intensities than do the lines of radical-ions that are not in a degenerate ground state (cf. Section 1.3). In fact, such saturation properties have been observed experimentally[46,205].

In the radical-anion of the substituted perimeter, the additional unpaired electron occupies that of the two lowest antibonding HMO's, ψ_{a+} or ψ_{a-}, which has the lower energy. The radical-cation, on the other hand, is formed by the removal of an electron from that of the two highest bonding HMO's, ψ_{b+} or ψ_{b-}, which has the higher energy, so that an unpaired electron is left in this HMO. It should therefore be possible to tell from the π-spin population whether the unpaired electron is delocalized according to ψ^2_{a+} or ψ^2_{a-} in the anion, and according to ψ^2_{b+} or ψ^2_{b-} in the cation. The ESR spectra of such radical-ions should thus give information about the relative energies of the corresponding HMO's in the substituted perimeter. This information will allow conclusions to be drawn on the effect that the substituents exert on the HMO's of the original pure perimeter. Some examples will show how the relevant experimental data can be interpreted in this way [209].

Six-membered ring. The simplest representative of neutral (4r + 2) perimeters is benzene (r = 1; 2m = 6). Its radical-anion displays a hyperfine structure that agrees with the expected uniform spin distribution (cf. Section 1.4, Figure 10). Figure 33 shows schematically the lowest degenerate antibonding HMO's

$$\psi_{a+} = 0.577\,(\Phi_1 + \Phi_4) - 0.289\,(\Phi_2 + \Phi_3 + \Phi_5 + \Phi_6).$$

and

$$\psi_{a-} = 0.500\,(\Phi_2 - \Phi_3 + \Phi_5 - \Phi_6).$$

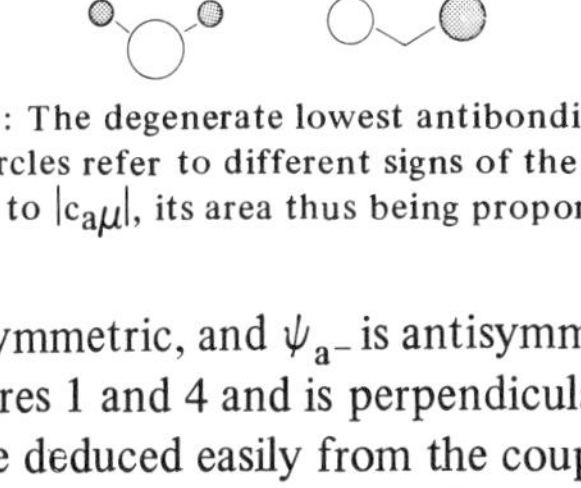

Figure 33: The degenerate lowest antibonding HMO's ψ_{a+} and ψ_{a-} of benzene. The blank and the shaded circles refer to different signs of the LCAO-coefficients $c_{a\mu}$. The radius of a circle is proportional to $|c_{a\mu}|$, its area thus being proportional to $c^2_{a\mu}$.

ψ_{a+} is symmetric, and ψ_{a-} is antisymmetric with respect to a mirror plane that traverses the centres 1 and 4 and is perpendicular to the plane of the ring.
It can be deduced easily from the coupling constants of the radical-ions of toluene and the three isomeric xylenes (LXXVIII-LXXXI) [188,190] that the spin population is distributed according to ψ^2_{a-} in LXXVIII$^{\ominus}$ and LXXIX$^{\ominus}$, but according to ψ^2_{a+} in LXXX$^{\ominus}$ and LXXXI$^{\ominus}$.

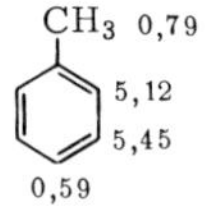

CH_3 0,79; 5,12; 5,45; 0,59
LXXVIII$^{\ominus}$

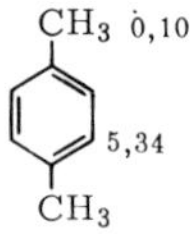

CH_3 0,10; 5,34; CH_3
LXXIX$^{\ominus}$

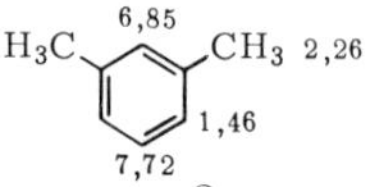

H_3C 6,85 CH_3 2,26; 1,46; 7,72
LXXX$^{\ominus}$

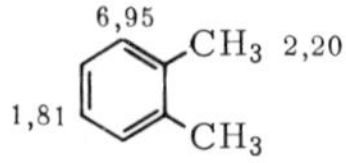

6,95 CH_3 2,20; 1,81; CH_3
LXXXI$^{\ominus}$

This result is explained satisfactorily by the weak inductive effect of the methyl substituents.

The HMO's ψ_{a+} and ψ_{a-} in benzene are stabilized (their energy lowered) or destabilized (their energy raised) by substituents, according to whether these are electron-attracting or electron-repelling. The extent of this energy shift depends primarily on $\sum_{\tilde{\mu}} c^2_{a\tilde{\mu}}$, where $c^2_{a\tilde{\mu}}$ are the squares of the LCAO coefficients at the substituted centres $\tilde{\mu}$[28,211].

The following positions of substituents are consistent with the symmetryof the degenerate HMO's ψ_{a+} and ψ_{a-} (cf. property 2 of perimeter HMO's and Figure 33):

1 (or 4) on monosubstitution ,
1 and 4 on p-disubstitution ,
2 and 6 (or 3 and 5) on m-disubstitution and
2 and 3 (or 5 and 6) on o-disubstitution.

It can be seen from Figure 34 that in the case of monosubstitution and 1,4-disubstitution, ψ_{a+} is strongly perturbed ($c^2_{a+,1} = c^2_{a+,4} = 0.333$), while ψ_{a-}, whose nodal plane traverses the centres $\tilde{\mu}$ = 1 and 4, remains unaffected ($c^2_{a-,1} = c^2_{a-,4} = 0$). In the case of 2,6- and 2,3-disubstitution, on the other hand, the perturbation of ψ_{a-} is stronger than that of ψ_{a+} ($c^2_{a+,2} = c^2_{a+,3} = c^2_{a+,5} = c^2_{a+,6} = 0.083$; $c^2_{a-,2} = c^2_{a-,3} = c^2_{a-,5} = c^2_{a-,6} = 0.250$).

It is therefore to be expected that the levels of ψ_{a+} and ψ_{a-} will be modified according to the direction of the inductive effect and the position of the substituents, as shown in Figure 34.

Since alkyl groups repel electrons, the unpaired electron in LXXVIII$^{\ominus}$ and LXXIX$^{\ominus}$ should occupy the antisymmetric HMO ψ_{a-} ; in LXXX$^{\ominus}$ and LXXXI$^{\ominus}$, for the same reason, it should occupy the symmetric HMO ψ_{a+} . This is exactly what has been deduced from the experimental data.

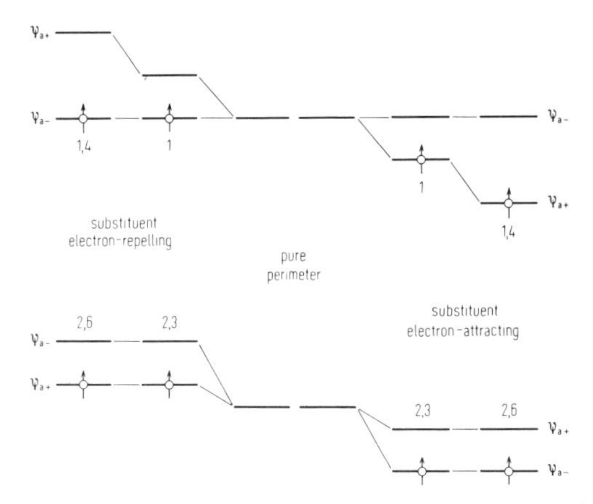

Figure 34: Splitting of the degenerate lowest antibonding HMO's ψ_{a+} and ψ_{a-} of benzene as a result of substitution, and their occupancy in the radical-anions of substituted derivatives. The numbers refer to the positions μ of the substituents (cf. Figure 33).

One should be able, of course, to check the prediction derived from Figure 34 for electron-attracting substituents as well. Thus, comparison between experimental data for LXXVIII$^{\ominus}$- LXXXI$^{\ominus}$and those for the corresponding cyano-derivatives LXXIV$^{\ominus}$- LXXVII$^{\ominus}$shows that the singly occupied HMO's in the two substituted series differ in symmetry: In the radical-anions LXXIV$^{\ominus}$and LXXV$^{\ominus}$of benzonitrile and terephthalonitrile, respectively, the spin population is distributed approximately according to ψ^2_{a+}; on the other hand, in the radical-anions LXXVI$^{\ominus}$ and LXXVII$^{\ominus}$ of isophthalonitrile and phthalonitrile, respectively, the spin distribution is more in accordance with ψ^2_{a-}. However, the agreement between experimental data and the values based on the perimeter model is poorer here than in the case of methyl-derivatives, since the requirement that the substituent should exert only a weak and essentially inductive perturbation on the HMO's of benzene is not fulfilled for the cyano group.

Ten-membered ring. The next representative of neutral (4r + 2)-perimeters, in which r =2 and 2m=10, cyclodecapentaene,is unknown.*) However, its derivative with a 1,6-methylene bridge (LXXXII) has recently been prepared. The physical and chemical properties of this bridged derivative indicate extensive π-electron delocalization[212a,325], despite the considerable non-planarity of the peripheral ring[326].

*) There is some evidence of cyclodecapentaene being a transient intermediate in photochemical reactions[329].

The HMO's of LXXII can be correlated with the HMO's of the pure ten-membered perimeter, as was done in the case of the alkyl-derivatives of benzene. Figure 35 shows schematically the lowest degenerate antibonding HMO's of the hypothetical cyclodecapentaene:

$$\psi_{a+} = 0.447\,(\Phi_1 - \Phi_6) - 0.138\,(\Phi_2 - \Phi_5 - \Phi_7 + \Phi_{10}) - 0.362\,(\Phi_3 - \Phi_4 - \Phi_8 + \Phi_9)$$

and

$$\psi_{a-} = 0.425\,(\Phi_2 + \Phi_5 - \Phi_7 - \Phi_{10}) - 0.263\,(\Phi_3 + \Phi_4 - \Phi_8 - \Phi_{10})$$

The perimeter is drawn in Figure 35 as a flattened ten-membered ring of LXXXII. This form of a perimeter can also be obtained by distorting the regular polygon in such a way that all the centres μ remain in one plane. Since the HMO's are insensitive to such deformations, the degeneracy of ψ_{a+} and ψ_{a-} is retained.

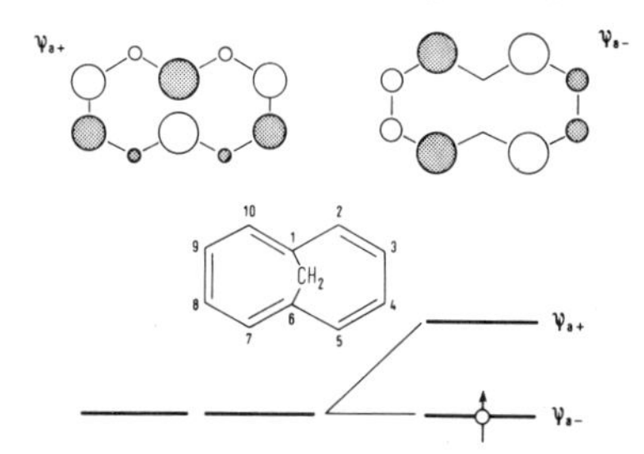

Figure 35: Degenerate lowest antibonding HMO's ψ_{a+} and ψ_{a-} of the hypothetical cyclodecapentaene (schematic representation as in Figure 33).
Bottom splitting of the HMO's ψ_{a+} and ψ_{a-} in 1,6-methano-clodecapentaene and their occupancy in the radical-anion.

Of the two orbitals, ψ_{a+} is symmetric and ψ_{a-} is antisymmetric with respect to the plane that passes through the substituted centres 1 and 6. Figure 35 shows that the inductive effect of the methylene bridge between these centres should not influence the HMO ψ_{a-} ($c^2_{a-,1} = c^2_{a-,6} = 0$). On the other hand, one expects that the HMO ψ_{a+} will be strongly perturbed ($c^2_{a+,1} = c^2_{a+,6} = 0.200$). If the inductive effect of the methylene bridge in LXXXII operates in the same direction as that of the methyl groups in the benzene derivatives LXXVIII-LXXXI, then ψ_{a-} must again lie lower than ψ_{a+} (cf. Figure 35). This statement is confirmed by experimental data for the radical-anion of LXXXII, since the observed values[97] are compatible with the distribution ψ^2_{a-}, but not with ψ^2_{a+}.

2,71
CH_2 1,14
0,10

LXXXII$^{\ominus}$

It is interesting to note that the experimental coupling constants of the ring protons (2.71 and 0.10 gauss) are much smaller than expected on the basis of the calculated spin populations and eq. (20). The discrepancy may be due to the above-mentioned lack of planarity of the peripheral ring in $LXXXII^{\ominus}$. This non-planarity may lead to a finite overlap between the π-electron system and the σ-orbitals of the C-H bonds, and to a direct delocalization of the unpaired electron into the ls atomic orbital of the hydrogens in the ring, as in the case of the β-protons of alkyl substituents (cf. Section 1.5). Such delocalization must reduce the effective coupling constants $a_{H\mu}$, because the additional ls-spin population at the ring protons has the same sign as ρ_{μ}^{π}, and thus it partly makes up for the spin population with an opposite sign resulting from σ-π spin polarization[97].

Fourteen-membered ring. Trans-15, 16-dimethyldihydropyrene (LXXXIII)[213] is an alkyl-bridged derivative of cyclotetradecaheptaene, the third representative of neutral (4r + 2)-perimeters (r = 3 and 2m = 14). Cyclotetradecaheptaene itself has been known for some years as an unstable [14] annulene lacking planarity for steric reasons[214]. Attempts to prepare its radical-ions have failed so far[215]. On the other hand, it is easy to produce[54] both the radical-anion and the radical-cation from the alkyl-derivative LXXXIII, whose π-electron centres lie virtually in one and the same plane [216]. The ESR studies of these radical-ions make it possible to check experimentally the theoretical predictions for the inductive effect of the alkyl bridge in LXXXIII on the four HMO's under consideration:

$$\begin{aligned}
\psi_{a+} &= -0.084\,(\Phi_1 + \Phi_3 + \Phi_6 + \Phi_8) + 0.378\,(\Phi_2 + \Phi_7) \\
&\quad + 0.236\,(\Phi_4 + \Phi_5 + \Phi_9 + \Phi_{10}) - 0.340\,(\Phi_{11} + \Phi_{12} + \Phi_{13} + \Phi_{14});\\
\psi_{a-} &= -0.368\,(\Phi_1 - \Phi_3 + \Phi_6 - \Phi_8) - 0.296\,(\Phi_4 - \Phi_5 + \Phi_9 - \Phi_{10}) \\
&\quad + 0.164\,(\Phi_{11} - \Phi_{12} + \Phi_{13} - \Phi_{14});\\
\psi_{b+} &= 0.084\,(\Phi_1 + \Phi_3 - \Phi_6 - \Phi_8) + 0.378\,(\Phi_2 - \Phi_7) \\
&\quad - 0.236\,(\Phi_4 - \Phi_5 - \Phi_9 + \Phi_{10}) - 0.340\,(\Phi_{11} + \Phi_{12} - \Phi_{13} - \Phi_{14});
\end{aligned}$$

and

$$\begin{aligned}
\psi_{b-} &= 0.368\,(\Phi_1 - \Phi_3 - \Phi_6 + \Phi_8) + 0.296\,(\Phi_4 + \Phi_5 - \Phi_9 - \Phi_{10}) \\
&\quad + 0.164\,(\Phi_{11} - \Phi_{12} - \Phi_{13} + \Phi_{14}).
\end{aligned}$$

Figure 36 shows the four HMO's, The perimeter was fitted to the planar form of the fourteen-membered ring of LXXXIII. As mentioned before, the degeneracy of the HMO's is retained with this form of the perimeter.

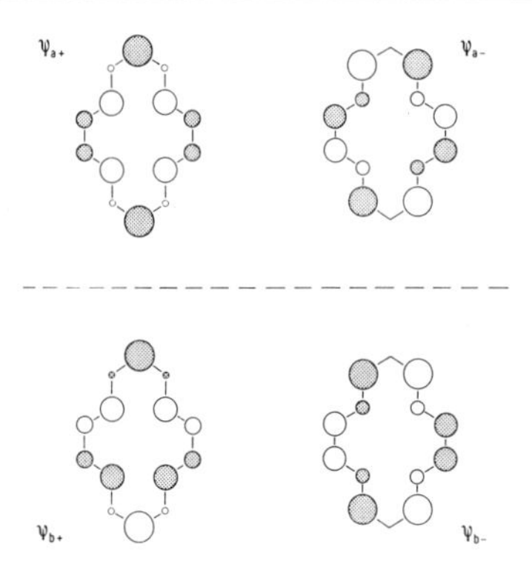

Figure 36: Degenerate lowest antibonding HMO's ψ_{a+} and ψ_{a-}, together with the degenerate highest bonding HMO's ψ_{b+} and ψ_{b-} of the hypothetical cyclotetradecaheptaene (schematic representation as in Figure 33).

The HMO's ψ_{a+} and ψ_{b+} are symmetric, while the HMO's ψ_{a-} and ψ_{b-} are antisymmetric with respect to the plane passing through centres 2 and 7. In accordance with the pairing properties of alternant perimeters, the following equations hold:

$$c^2_{a+,\mu} = c^2_{b+,\mu} \quad \text{and} \quad c^2_{a-,\mu} = c^2_{b-,\mu}. \tag{128}$$

Considerations analogous to those made for the alkyl-substituted derivatives of benzene and cyclodecapentaene suggest that the inductive effect of the alkyl bridge in LXXXIII should destabilize the symmetric HMO's ψ_{a+} and ψ_{b+} more strongly than the antisymmetric HMO's ψ_{a-} and ψ_{b-} ($c^2_{j+,\mu} = 0.116 > c^2_{j-,\mu} = 0.027$, where j = a or b; and μ = 11,12,13,14). It is therefore to be expected that ψ_{a-} will be of lower energy than ψ_{a+}. Furthermore, ψ_{b-} will be more stable than ψ_{b+} (cf. Figures 36 and 37). The unpaired electron should consequently occupy the antisymmetric orbital ψ_{a-} in the radical-anion and the symmetric orbital ψ_{b+} in the radical-cation. However, the experimental data indicate that the spin population in both radical-ions is distributed according to the squares of the symmetrical functions[54], i.e. ψ^2_{a+} in the anion and ψ^2_{b+} in the cation.

5,46 CH₃ 0,20 0,78 16 15 CH₃ 0,78

LXXXIII$^{\ominus}$

4,78 CH₃ 0,10 1,03 CH₃ 1,50

LXXXIII$^{\oplus}$

Thus, the assumption that the central bridge in LXXXIII exerts a purely inductive effect leads to agreement with experimental data only in the case of the radical-cation. In the case of the radical-anion, however, the observed values contradict the theoretical deductions. To find a plausible explanation for this conflict, one should consider the geometry of the alkyl bridge in LXXXIII. Since the methyl groups of the bridge lie above and below the plane of the fourteen - membered ring, the σ-bonds between C15 or C16 and the methyl carbons are virtually perpendicular to this plane[213,216]. It is probable that the hyperconjugation of these σ-bonds with the π-electron system of the perimeter plays an important part. As Figure 37 shows, full agreement with experimental data can be reached by considering such a hyperconjugative effect superimposed on a purely inductive one[54].

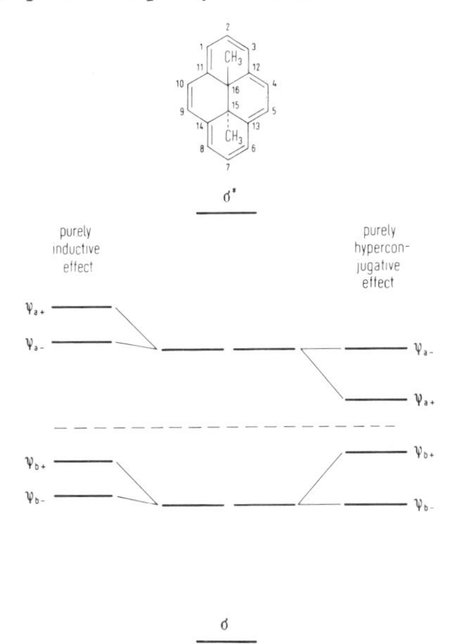

Figure 37: Splitting of the degenerate lowest antibonding HMO's ψ_{a+} and ψ_{a-} and the degenerate highest bonding HMO's ψ_{b+} and ψ_{b-} of cyclodecaheptaene, brought about by an inductive (left) and a hyperconjugative effect (right) of the central alkyl bridge in 15,16-dimethyldihydropyrene.

4r-Perimeters: Eight-membered ring. It has been already mentioned that the MO models of perimeters with 2m = 4r centres differ from those of perimeters with 2m = 4r + 2 centres by the presence of degenerate non-bonding HMO's ψ_{n+} and ψ_{n-}. In the ground state of a neutral 4r-perimeter, the bonding HMO's are occupied by (2m - 2)

π-electrons, while the remaining two electrons must be placed into the non-bonding orbitals ψ_{n+} and ψ_{n-}. An unstable triplet ground state,in which each of the degenerate HMO's ψ_{n+} and ψ_{n-} accommodates one electron, is therefore to be expected for such a perimeter (cf. Figure 38). The normal ground state of an aromatic system, the singlet state, should become possible only for the di-anion or the di-cation of a 4r-perimeter, in which the HMO's ψ_{n+} and ψ_{n-} are both filled by the acquisition of two π-electrons, or become vacant by the release ot two π-electrons. According to expectation, there exists no neutral eight-membered perimeter with a fully conjugated π-electron system. The singlet ground state cyclooctatetraene (XII), which has long been known possesses a non-planar boat form[217] and resembles olefins in its properties[218]. On the other hand, its dianion $XII^{\ominus\ominus}$ exhibits a planar arrangement of the eight carbon centres, together with a delocalization of the π-electrons[219]. Reduction of cyclooctatetraene to the dianion $XII^{\ominus\ominus}$ is preceded by the formation of a paramagnetic intermediate, the radical-anion $XII^{\ominus}$, whose nine-line ESR spectrum points to interaction with eight equivalent protons[155,193a]. The coupling constant $a_{H\mu} = 3.21$ gauss is consistent with the spin population $\rho_{\mu}^{\pi} = 0.125$ calculated for each centre of a planar eight-membered ring. Other results also suggest that the radical-anion $XII^{\ominus}$ has a planar eight-membered perimeter, similar to the diamagnetic dianion $XII^{\ominus\ominus}$[155,219].

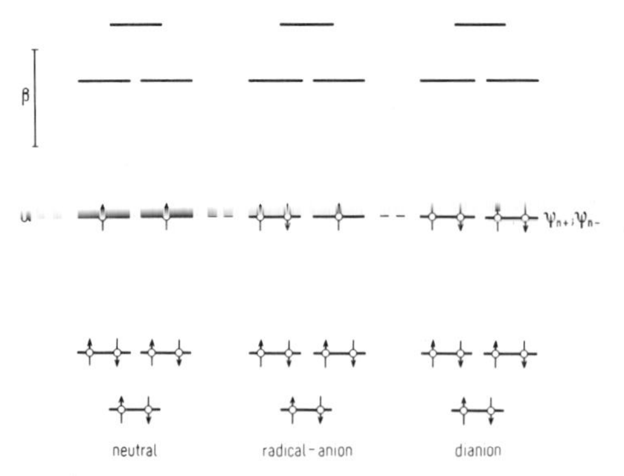

Figure 38: HMO energy diagram of a 4r-perimeter (eight-membered ring). Occupancy in the neutral system, the radical-anion, and the dianion.

In the radical-anion $XII^{\ominus}$, the degenerate non-bonding HMO's ψ_{n+} and ψ_{n-} are occupied by three π-electrons, of which two are paired and one is unpaired. This represents an essential difference from the situation in the radical-anions of (4r + 2)-perimeters, in which one unpaired electron has to choose between two antibonding HMO's ψ_{a+} and ψ_{a-}. Nevertheless, since there are two equi-energetic configurations in $XII^{\ominus}$ (doubly occupied ψ_{n+} with singly occupied ψ_{n-} and vice versa), the ground state should be degenerate, as in the radical-anions of the (4r + 2) perimeters. This degeneracy is lifted by the introduction of an alkyl substituent into the eight membered ring. Figure 39 is a schematic representation of the degenerate non-bonding HMO's

$$\psi_{n+} = 0.500\,(\Phi_1 - \Phi_3 + \Phi_5 - \Phi_7)$$

and

$$\psi_{n-} = 0.500\,(\Phi_2 - \Phi_4 + \Phi_6 - \Phi_8)$$

of planar cyclooctatetraene; ψ_{n+} is symmetric and ψ_{n-} antisymmetric with respect to a mirror plane passing through centres 1 and 5. If one of these centres carries an alkyl group, the inductive effect of the latter is expected to destabilize the HMO ψ_{n+} ($c^2_{n+,1}$ = 0.250), while the energy of ψ_{n-} will remain unchanged ($c^2_{n-,1}$ = 0). Thus, in the radical-anion of monoalkyl-cyclooctatetraene that configuration will be favoured in which ψ_{n-} is doubly occupied and ψ_{n+} is singly occupied, i.e. the unpaired electron spin should have the distribution ψ^2_{n+} (cf. Figure 39).

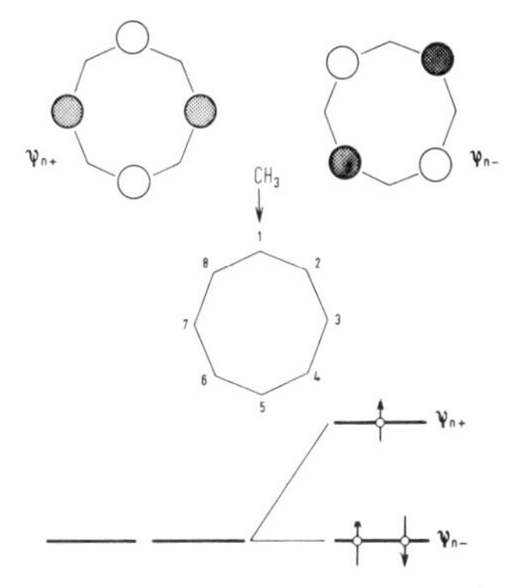

Figure 39: Degenerate non-bonding HMO's ψ_{n+} of planar cyclooctatetraene (schematic representation as in Figure 33).
Bottom: splitting of the HMO's ψ_{n+} and ψ_{n-} in the alkyl-substituted planar cyclooctatetraene and their occupancy in the radical-anion.

The experimental data [193] for the radical-anion of methylcyclooctatetraene are given in the formula LXXXIV$^{\ominus}$, together with the values (in brackets) of the coupling constants $a_{H\mu}$ obtained on the basis of eq. (20) for the two cases in which either ψ_{n+} or ψ_{n-} is occupied. (The value used for $|Q_{CH}|$ in eq. (20) was the overall range A = 8 x 3.21 = 25.68 gauss, measured in the ESR spectrum of the unsubstituted radical-anion XII$^{\ominus}$).

CH$_3$ 5,1 1,6 4,8 1,6 4,8

LXXXIV$^{\ominus}$

Ψ^2_{n+}	Ψ^2_{n-}
(0 ;	6,4)
(6,4 ;	0)

Comparison of the calculated and the observed coupling constants indicates that the spin population in LXXXIV$^{\ominus}$ follows the distribution ψ^2_{n+} rather than ψ^2_{n-}. However, the HMO ψ_{n+} cannot be solely responsible for the experimental data, and a certain amount of participation of the HMO ψ_{n-} is necessary to obtain better agreement between theory and experiment. In other words, there must be a certain mixing between the two HMO's that are degenerate in the pure perimeter. This mixing, which can also be assumed for other substituted perimeters considered so far, will be discussed below.

Orbital mixing. The splitting between two degenerate orbitals that arises on introductio of an alkyl substituent into the perimeter widens with the extent of the perturbation. The spin population in the substituted radical-anion is then determined by an increasing contribution of the orbital with the lower energy, and, consequently departs more and more from the uniform spin distribution in the radical-anion of an unperturbed perimeter. This spin population can be expressed by the sum of contribution of the two orbitals in question[189,209,231]. It may be written as

$$\psi_a^2 = C_+^2\,\psi_{a+}^2 + C_-^2\,\psi_{a-}^2 \quad \text{for } (4r+2)\text{-perimeter}$$

or

$$\psi_n^2 = C_+^2\,\psi_{n+}^2 + C_-^2\,\psi_{n-}^2 \quad \text{for } 4r\text{-perimeter} \qquad (129)$$

$$(C_+^2 + C_-^2 = 1)$$

where the squared coefficients C_+^2 and C_-^2 are a measure of the participation of the symmetrical (ψ_{a+} or ψ_{n+}) and the antisymmetrical (ψ_{a-} or ψ_{n-}) orbitals, respectively. In the case of the radical-anions of substituted benzene derivatives, this participation has been discussed by several authors[2,189,209]. For the radical-anion of benzene itself (I$^{\ominus}$), the relationship $|C_+| = |C_-| = 1/\sqrt{2}$ should clearly apply. For the alkyl derivatives of benzene, on the other hand, one might expect that either of two limiting cases of eq. (129) will hold according to the position of the substitution: $C_+ = 0$ and $|C_-| = 1$; or $|C_+| = 1$ and $C_- = 0$. This has been assumed so far in the case of the methyl-substituted radical-anions LXXVIII$^{\ominus}$ - LXXXI$^{\ominus}$. In actual fact, however, one does not deal here with limiting cases, but with the relationships:

$|C_+| < |C_-|$ in $LXXVIII^{\ominus}$ and $LXXIX^{\ominus}$, and

$|C_+| > |C_-|$ in $LXXX^{\ominus}$ and $LXXXI^{\ominus}$.

The following values have been estimated for the radical-anions $LXXVIII^{\ominus}$ (toluene) and $LXXIX^{\ominus}$ (p-xylene) by the combination of experimental results (ESR and proton resonance spectra) with MO calculations (CI-method)[220,231]:

$LXXVIII^{\ominus}$: $|C_+| = 0.50$ and $|C_-| = 0.87$;

$LXXIX^{\ominus}$: $|C_+| = 0.35$ and $|C_-| = 0.94$.

There are two ways in which the orbital of higher energy can mix with the lower one and contribute to the spin population in the radical-anion. Firstly, the Boltzmann distribution law requires some participation of the higher orbital (cf. eq. (6) in Section 1.1), provided the orbital splitting is not too large (thermal mixing). Secondly, the coupling of the two orbitals through vibrations of appropriate symmetry must be considered (vibronic mixing). This coupling, which will not be formally derived here, is thought to be by far the more important mechanism of orbitai mixing in most cases[209,231].

The ESR spectra of the radical-anions of partially deuterated perimeters are of great interest with regard to orbital mixing. Two such systems have been investigated: the radical-anion of monodeuterated benzene (LXXXV) and that of monodeuterated cyclooctatetraene (LXXXVI). In the case of $LXXXV^{\ominus}$[221], the introduction of deuterium into the ring results in a splitting of the degenerate orbitals ψ_{a+} and ψ_{a-}. Owing to the extremely weak effect of the substituent, however, this splitting is very small, so that a particularly intense mixing of the orbitals ψ_{a+} and ψ_{a-} comes about.

It follows from the experimental data given in the formula ($LXXXV^{\ominus}$) that the contribution of ψ_{a-} to the spin population is somewhat greater than that of ψ_{a+}. This result agrees with the generally accepted view according to which the secondary isotope effect exerted by deuterium on a π-electron system is electron-repelling, and thus has the same direction as the inductive effect of alkyl substituents[189,222].

D 0,55 (3,58) *)
3,92
3,92
3,41

$LXXXV^{\ominus}$

D 0,50 (3,25) *)
3,2
3,2
3,2
3,2

$LXXXVI^{\ominus}$

*) The values in parentheses refer to the coupling constants of deuterons multiplied by the factor $g_N(H)/g_N(D) = 1/0.1535$ (cf. Section 1.4).

Unlike the spectrum of LXXXV$^{\ominus}$, that of the radical-anion LXXXVI$^{\ominus}$ of mono-deuterated cyclooctatetraene is compatible with a uniform spin distribution [223]. The two non-bonding orbitals ψ_{n+} and ψ_{n-} thus seem to remain degenerate in spite of the introduction of deuterium, and to make equal contributions to the spin population. This unexpected result can be rationalized by the symmetry relationships of ψ_{n+} and ψ_{n-} [209,223], but a different interpretation has also been proposed [224,328].

2.6. Non-aromatic systems

This Section deals with radical-ions of systems in which the cyclic delocalization of π-electrons is limited or virtually non-existent. Such radical-ions are formed from ethylene derivatives, polyenes, radialenes, and 1,2-diones. Nitroalkanes are also included here, since the nitro group can be treated as a π-electron system with three hetero-centres.

Only a few reports have been published about radical-ions of organic compounds containing no π-electrons. Thus, it has been reported that radical-anions were prepared from cyclopropane [225], adamantane [226], and hexamethylenetetraamine [226]. However, the ESR spectrum attributed to the adamantane radical-anion is identical with the spectrum of the radical-anion of benzene [227], and is probably due to benzene present as an impurity in the solvent [228]. The existence of the radical-anions of hexamethylene-tetramine and cyclopropane has been also impossible to confirm, since the spectra first obtained for these species could not be reproduced [228]. The identity of the radical-cation of trimethylenediamine, on the other hand, appears to be beyond doubt in ESR spectroscopic work [285].

Ethylene derivatives. Tetracyanoethylene (LXXXVII), which contains four electron acceptor groups, has a sufficiently high electron affinity to give a very stable radical-anion [70,229,230]. Analogously, tetrakis (dimethylamino) ethylene (LXXXVIII) possesses four electron-donor groups and has such a low ionization potential that it can be easily converted into the radical-cation [76].

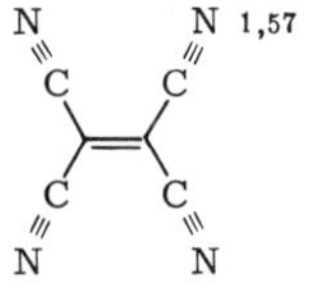

N 1,57

LXXXVII

H_3C CH_3 3,28 or

H_3C-N 4,85 N-CH_3 2,84

H_3C-N N-CH_3

H_3C CH_3

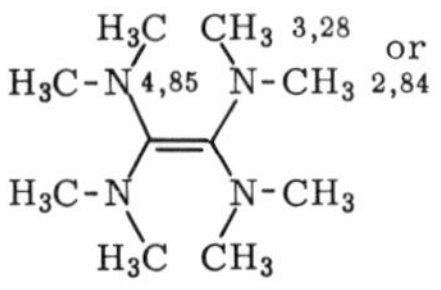

LXXXVIII$^{\oplus}$

By means of eqs (127) and (122) one can compare the coupling constants of the ^{14}N nuclei in LXXXVII$^{\ominus}$(a_N^{CN}) and in LXXXVIII$^{\oplus}$ ($a_N^{N(CH_3)_2}$) with the calculated spin populations[68,76]. The same or similar HMO parameters α_N and β_{CN} which have been used with the correspondingly substituted aromatic radical-ions[68,77] (cf. Section 2.4) are also adequate in these calculations.

The two different values given with the formula LXXXVIII$^{\ominus}$ for the coupling constant of the methyl protons in an $N(CH_3)_2$ substituent indicate that the two methyl groups of the substituent are not equivalent. This points to hindrance of free rotation about the formally single C-N bond.

Conjugated polyenes. The preparation of the radical-anion of trans-1,3-butadiene (LXXXIX) has made one of the simplest electron systems accessible to ESR investigations[73].

H H 7,62
H H 7,62
H H
2,79

LXXXIX$^{\ominus}$

Comparison of the coupling constants a_{H1} and a_{H2} with the HMO values $c_{a1}^2 = 0.362$ and $c_{a2}^2 = 0.138$ shows that the simplest HMO model (in which all bond integrals β have the same magnitude) provides a good agreement between experiment and theory in LXXXIX$^{\ominus}$. The proportionality factor $|Q_{CH}|$ has an average value of 20.8 gauss, which is smaller than that found for most aromatic radical-anions (cf. Section 2.1). Within the limits of the resolution achieved, the four terminal protons form an equivalent set.

Refined MO methods give poorer agreement with experiment than does the simple HMO model[73]. It can be assumed here, as in many similar cases, that the surprising success of this model is due to the fact that the effects of numerous simplifications cancel.

In compound XC, two cis-1,3-butadiene systems are separated by sp^3-hybridized carbons[212b].

H_2
C 3,80
O
1,51
C
H_2 0,45

XC$^{\ominus}$

The hyperfine structure of the radical-anion $XC^{\ominus}$ due to three sets of four equivalent protons, indicates[97] that the spin populations is equally shared between the two butadiene fragments of the molecule. The coupling constants of the ring protons are about half the values found for the radical-anion of trans-1,3-butadiene (cf. formulae $LXXXIX^{\ominus}$ and $XC^{\ominus}$).

Radical-anions of other compounds containing the 1,3-diene system have been the subject of a recent report[284]. Furthermore, a study has been described of the radical-anion of 1,3,5-cycloheptatriene (XCI)[232] whose π-electron system is very similar to that of an all-cis-1,3,5-hexatriene.

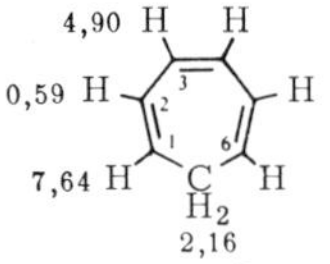

4,90 H H
0,59 H H
7,64 H C H
H_2
2,16
$XCI^{\ominus}$

The HMO values $c_{a1}^2 = 0.272$, $c_{a2}^2 = 0.054$ and $c_{a3}^2 = 0.175$ calculated for 1,3,5-hexatriene closely approximate the spin populations of $XCI^{\ominus}$, as can be seen by comparing these values with the coupling constants a_{H1}, a_{H2} and a_{H3}. The agreement is further improved by using McLachlan's equation (44) and suitable HMO parameters to allow for the effect of the methylene group and the deviation of the π-electron system from coplanarity[232]. The magnitude of the coupling constant $a_H^{CH_2}$ of the methylene protons in $XCI^{\ominus}$ is of particular interest. Since the lowest vacant HMO ψ_a in 1,3,5-hexatriene possesses a nodal plane between centres 1 and 6, the hyperconjugative HMO model leads to vanishing spin population in the methylene group. Even the McLachlan method yields only a fraction of the spin population that is compatible with the observed coupling constant $a_H^{CH_2} = 2.16$ gauss. It is assumed that in this special case the most important role is played by the spin polarisation between the σ-electrons of the C-H bonds of the methylene group and the unpaired π-electron at centres 1 and 6[124,232]. The extent of such σ-π polarization should be a whole order of magnitude smaller than in the case of ring protons (α) (cf. Section 1.5), since the methylene protons (β) are separated from the π-electron system by two σ-bonds. This is the reason why in radical-ions in which hyperconjugation can be fully effective, the contribution made by the spin polarization to the coupling constants of β-alkyl protons is usually neglected[123].

[n] **Radialenes.** These are $C_{2n}H_{2n}$ hydrocarbons containing n radially disposed double bonds. The known members of this group are the unsubstituted [3] and [4] radialenes[233], hexamethyl[3]radialene[234] (XCII), and the two hexaalkyl-derivates XCIII

($R = CH_3$ or C_2H_5) of [6]radialene[235]. It has been possible to prepare[236,237] the radical-anions of the three alkyl-substituted compounds. The experimental data are given below for the radical-anions $XCII^{\ominus}$ and $XCIII^{\ominus}$ ($R = CH_3$) of hexamethyl[3]-radialene and 1,2,3,4,5,6-hexamethyl[6]radialene, respectively.

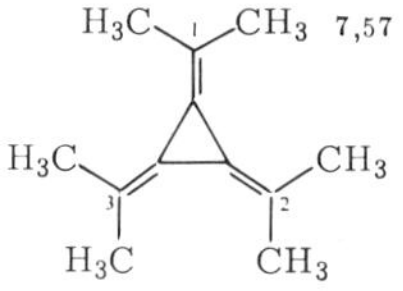

$XCII^{\ominus}$ (CH_3: 7,57)

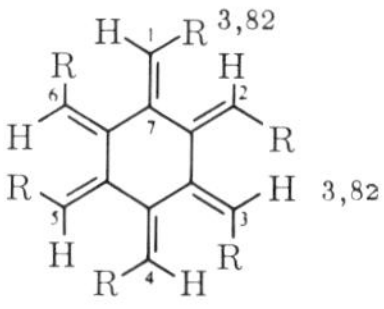

$XCIII^{\ominus}$ ($R = CH_3$) (3,82; H 3,82)

Theoretical and experimental studies of neutral compounds XCII and XCIII indicate that the delocalization over the three- or six-membered ring, i.e. between the radially disposed double bonds, is very slight in the ground state of the radialenes. This should be particularly so with the hexaalkyl-derivatives XCIII, in which steric hindrance makes it impossible for the system to have a planar arrangement of the carbon atoms in the six-membered ring[238]. Going from the neutral compounds XCII and XCIII to their radical-anions, one expects no significant change in this respect, since according to HMO models, uptake of an extra π-electron affects the bond order only slightly. All the more surprising is the fact that the simple HMO model which ignores both the virtual localization of the double bonds and the effect of the alkyl substituents, gives good approximations for the spin populations in $XCII^{\ominus}$ and $XCIII^{\ominus}$. The ratio between the $c^2_{a\tilde{\mu}}$ values 0.285 and 0.142, calculated for the peripheral centres of [3] radialene and [6] radialene, respectively ($\tilde{\mu}$ = 1,2,3; and 1 through 6; cf. formulae above), is 2. The same ratio has been found (within the experimental error) between the coupling constants $a_H^{CH_3}$ of the methyl protons in $XCII^{\ominus}$ and in $XCIII^{\ominus}$ ($R = CH_3$), i.e. between 7.57 and 3.82 gauss, although the π-charges $q^{\pi}_{\tilde{\mu}}$ at the above-mentioned centres $\tilde{\mu}$ in the two radical anions are very different[237] (cf. Section 1.5). Once again the unexpectedly good agreement between the experimental data and the values calculated with the aid of the simplest HMO model may be explained by assuming a mutual cancellation of various effects exerted on the spin populations $\rho^{\pi}_{\tilde{\mu}}$ and their relationships with the coupling constants $a_H^{CH_3}$. Such effects are bond localization, influence of the alkyl groups, deviation from coplanarity, and dependence on charge $q^{\pi}_{\tilde{\mu}}$.

1,2-Diones. The radical-anions of 1,2-diones, known as semidione-anions, are the non-aromatic analogues of semiquinone-anions (cf. Section 2.3). They occur as paramagnetic intermediates in the reduction of 1,2-diones to ene-1,2-diols:

$$\text{R(C=O)–(C=O)R}^1 \underset{-e^{\ominus}}{\overset{+e^{\ominus}}{\rightleftharpoons}} [\text{R–C(:O)=C(O}^{\cdot})\text{–R}^1]^{\ominus} \underset{-e^{\ominus}}{\overset{+e^{\ominus}}{\rightleftharpoons}} [\text{R–C(:O)=C(O:)–R}^1]^{\ominus\ominus}$$

cis- or trans-dione — cis- or trans-semidione-anion — dianion of cis- or trans-ene-diol

The simplest semidione-anion derives from glyoxal (R = R[1] = H). The coupling constant a_H of the two equivalent protons in this radical-anion is 7.7 gauss, and thus corresponds to a spin population of $\rho_C^{\pi} \approx 0.3$ at each carbon centre[286]. It is not certain whether the configuration is cis or trans[287]. On the other hand, both configurations have been identified for some alkyl-substituted semidione-anions (R = R[1] = alkyl)[288]. The two forms give rise to different ESR spectra, because the coupling constants of the β-alkyl protons*) are greater in the cis- than in the trans-isomer. This result suggests that there is a greater deviation from coplanarity in the cis-isomers. The higher stability of the trans-isomers is manifested in the concentration ratio which ranges between 10 and 100. The experimental data obtained for the radical-anions of biacetyl (XCIV) are given below[288]

$:O(H_3C)C{=}C(O^{\cdot})\overset{\beta}{C}H_3$ 7,0 — trans-XCIV$^{\ominus}$

$H_3C(:O)C{=}C(O^{\cdot})\overset{\beta}{C}H_3$ 5,6 — cis-XCIV$^{\ominus}$

concentration ratio 20:1

Unlike the open-chain semidione-anions, the monocyclic semidione-anions with small to moderately large rings (n ⩽ 9) exist exclusively in the cis configuration. The experimental data measured at +25 °C for such anions are here summarized[289,324]:

$[\beta H_2C–(CH_2)_{n-4}–CH_2,\ :O–C{=}C–O^{\cdot}]^{\ominus}$ XCV$^{\ominus}$

n	$a_{H(a)}^{\beta\text{-}CH_2}$	$a_{H(e)}^{\beta\text{-}CH_2}$
5	13,12	13,12
6	9,82	9,82
7	6,70	1,97
9	4,92	2,46

*) According to the definition given in Section 1.5, alkyl protons are designated as α, β, γ, etc., when they are linked to a π-electron centre via 0, 1, 2, etc. sp^3-hybridized carbon atoms. In the case of semidiones, this designation does not agree with the chemical notation normally used for ketones.

The coupling constants of the methylene β-protons in $XCV^{\ominus}$ vary greatly with the size of the ring, since they are strongly geometry-dependent as required by the mechanism of hyperconjugation (cf. Section 1.5). The protons in the axial (a) and the equatorial (e) positions are magnetically equivalent or non-equivalent, according to whether the life-time of a conformation is short or long in comparison with the reciprocal of the difference: $\gamma_E[a_{H(a)}^{\beta\text{-}CH_2} - a_{H(e)}^{\beta\text{-}CH_2}]$ (cf. Appendix A. 2.3).

Semidione-anions also arise when atmospheric oxygen reacts with monoketones containing an HCOH, HCBr, or an HCH group adjacent to the carbonyl function [289-291]. Thus, either trans- or cis-decalone gives a mixture of two radical-anions which have been identified by ESR spectroscopy as $XCVI^{\ominus}$ and $XCVII^{\ominus}$, and as $XCVI^{\ominus}$ and $XCVIII^{\ominus}$ in the trans and the cis case, respectively.

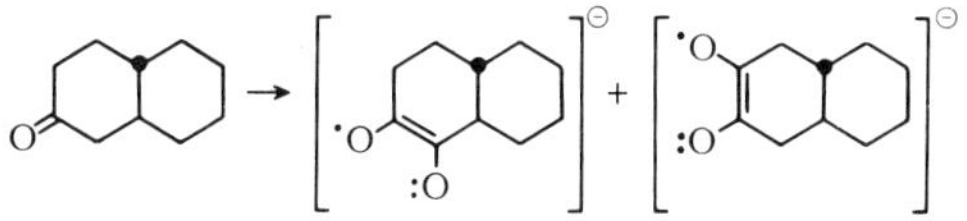

trans-2-decalone → $[XCVI]^{\ominus}$ + $[XCVII]^{\ominus}$

trans-2-decalone $XCVI^{\ominus}$ $XCVII^{\ominus}$

concentration ratio 3:2

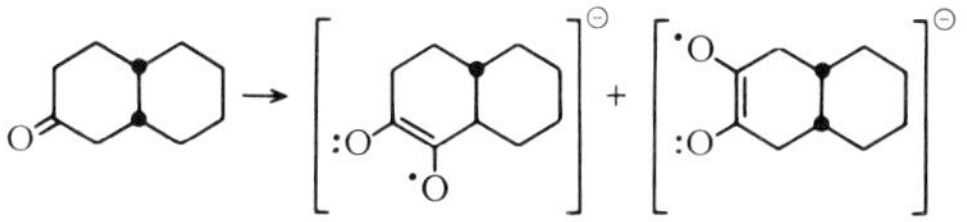

cis-2-decalone → $[XCVI]^{\ominus}$ + $[XCVIII]^{\ominus}$

cis-2-decalone $XCVI^{\ominus}$ $XCVIII^{\ominus}$

concentration ratio 1:3

The ESR spectra of the two semidione mixtures are very different and thus characteristic of the starting compounds. Similar observations have been made on the oxidation of other polycyclic ketones, and the configuration of the two corresponding six-membered rings in the ketones could therefore be determined. This result makes it possible to use ESR spectroscopy for the elucidation of the structures of sterones [291].

Aliphatic nitro compounds. The nitro group possesses such high electron affinity that it has been possible to prepare the radical-anions of many nitroalkanes and to investigate them by ESR spectroscopy [239]. The spectra differ in the proton hyperfine structure, according to whether the nitro substituent is attached to a primary, secondary, or tertiary carbon atom.

$$R\text{-}\underset{\underset{9{,}7\ \text{to}\ 10{,}1}{H}}{\overset{H}{C}}\text{-}NO_2 \quad 24{,}2 \text{ to } 25{,}5$$

XCIX$^{\ominus}$

$$R\text{-}\underset{\underset{3{,}2\ \text{to}\ 5{,}7}{H}}{\overset{R'}{C}}\text{-}NO_2 \quad 24{,}7 \text{ to } 25{,}7$$

C$^{\ominus}$

$$R\text{-}\underset{R''}{\overset{R'}{C}}\text{-}NO_2 \quad 23{,}8 \text{ to } 25{,}5$$

CI$^{\ominus}$

On the other hand, the coupling constants $a_N^{NO_2}$ of the ^{14}N nuclei vary only slightly with the type of the alkyl residue. They amount to 24.8 ± 1.0 gauss, and are suitable for checking the values $|Q_N| = 99 \pm 10$ and $|Q_{ON}| = 36 \pm 6$ gauss proposed[176] for the equation (cf. Section 2.4)

$$a_N^{NO_2} \approx Q_N \rho_N^\pi + 2 Q_{ON} \rho_O^\pi \qquad (121)$$

The π-spin populations at the nitrogen and the oxygens can be set equal to the squared LCAO coefficients calculated for the lowest vacant SCF-MO of nitromethane[240]:

$$\psi_a = 0.684\,\Phi_N - 0.514\,(\Phi_O - \Phi_{O'})$$

Substitution of these values $\rho_N^\pi = 0.468$, and $\rho_O^\pi = \rho_{O'}^\pi = 0.266$ into eq. (121) yields $|a_N^{NO_2}| = 27 \pm 8$ gauss, if different signs of Q_N and Q_{ON} are taken into account. The result of this calculation is thus in satisfactory agreement with the experimental value.

Trinitromethane has been the source of the doubly charged radical-anions CII$^{\ominus\ominus}$ and CIII$^{\ominus\ominus}$ (the latter can be prepared also from tetranitromethane) containing two and three nitro groups, respectively[241]:

$(O_2N)_2\dot{\dot{C}}$–H: 9,6 (N); H 4,1

CII$^{\ominus\ominus}$

$(O_2N)_2\dot{\dot{C}}$–NO_2: 8,4

CIII$^{\ominus\ominus}$

The interpretation of the ESR data of these two species is difficult, because there is no information concerning their geometry. If the carbon is assumed to be sp^2-hybridized, then CII$^{\ominus\ominus}$ and CIII$^{\ominus\ominus}$ can be treated as seven- and ten π-electron systems, respectively[294]. According to the HMO model, the π-spin populations should be zero at the carbon centre. McLachlan's formula (44) yields a negative spin population ρ_C^π at this centre, thus giving rise, by means of σ-π spin polarisation, to a positive contribution to the coupling constant $a_H = 4.1$ gauss of the adjacent proton in CII$^{\ominus\ominus}$.

A second and similarly positive contribution to a_H might come from hyperconjugation, if the π-electron centres in $CII^{\ominus\ominus}$ differ strongly from the coplanar arangement. The coupling constant $a_N^{NO_2}$ (= 8.4 gauss) for the trinitro-radical-dianion $CIII^{\ominus\ominus}$ is almost exactly a third of the values (24.8 ± 1.0 gauss) found for the mononitro-radical-anions $XCIX^{\ominus}$–$CI^{\ominus}$. The coupling constant $a_N^{NO_2}$ (= 9.6 gauss) for the dinitro-radical-dianion $CII^{\ominus\ominus}$, on the other hand, is considerably smaller than half this value.

Appendix to Part 2

A.2.1. ^{13}C Hyperfine structure

As mentioned earlier, the most abundant carbon isotope ^{12}C is non-magnetic (I=0), whereas the isotope ^{13}C with a natural abundance of 1.1% exhibits magnetic properties. In accordance with its spin quantum number of $I = \frac{1}{2}$, the ^{13}C nucleus gives rise to two hyperfine lines of equal intensity. Each line in the spectrum of an aromatic radical-ions is therefore accompanied by a pair or several pairs of weak satellites resulting from the small natural ^{13}C content of the sample. The twin satellites are situated at the same distance from the parent line due to ^{13}C-free radical-ions, and the intensity of each satellite relative to the parent line is (n/2) x 0.011, where n is the number of equivalent positions that can accommodate the ^{13}C nucleus. For this reason, ^{13}C satellites are found primarily when n is large, i.e. in the spectra of highly symmetric radical-ions[36,48,70,188,236,237,243-246]. Such satellites are conspicuous in the spectrum of the radical-anion XCIII$^{\ominus}$ of hexamethyl[6]radialene[236], shown in Figure 40.

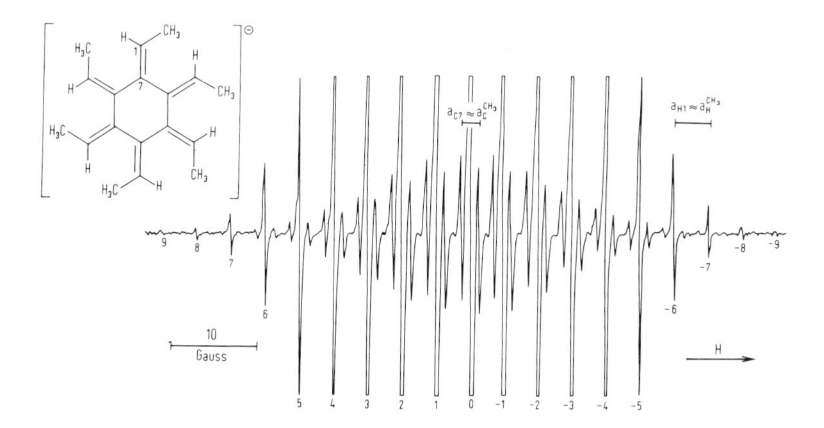

Figure 40: ESR spectrum of the radical-anion of 1,2,3,4,5,6-hexamethyl[6]radialene in 1,2-dimethoxyethane, at – 70 °C; gegenion $Na^{\oplus}$[236].
Each hyperfine line (–9 . . . . . 0, . . . +9) is flanked by two satellites, whose distance (2.00) gauss) is equal to the coupling constant $a_{C7} \approx a_C^{CH_3}$ of a ^{13}C nucleus. To make these ^{13}C satellites clearly recognizable, the signal must be considerably amplified, so that the most intense parent lines (–4, ... 0 . . . . +4) are off scale.

In many cases, however, all or nearly all weak ^{13}C satellites are hidden by the wings of the strong parent lines, so that they cannot be identified. The intensity of a satellite pair is increased by ^{13}C enrichment at one of the equivalent sites in question. This method has been applied by several authors [70,203,242,245,247,248].

The coupling constant $a_{C\mu}$ of a ^{13}C nucleus at centre μ is equal to the distance between the twin satellites, or twice the distance between a satellite and the parent line. Some relevant experimental data are listed in Table 13.

Table 13: Coupling constants of ^{13}C nuclei in some radical-ions.

Radical ion of	μ	$a^{\ominus}_{C\mu}$	$a^{\oplus}_{C\mu}$	^{13}C*)	Ref.
Benzene I	1	2,8		nat.	36)
Naphthalene II	1	7,3; 7,1		nat.; enr.	
	2	1,0; 1,2		nat.; enr.	242,243)
	9	5,6		nat.	
Anthracene III	1	3,57		nat.	
	2	0,25	0,37	nat.	
	9	8,76	8,48	enr.	153)
	11	4,59	4,50	nat.	
Biphenylene X	1	3,48	3,20	nat.	327)
	2	2,36	2,48	nat.	
Cyclooctatetraene XII		1,28		nat.	224)
1,4-Benzoquinone	1	0,40**)		enr	
XL	2	0,59		nat.	245)
Benzophenone XLIX	CO	9,13***)		enr.	248)
Benzonitrile LXXIV	CN	6,12; 6,4		enr.	70,203)
Terephthalonitrile	CN	7,83		enr.	
(1,4-Dicyanobenzene)	1	8,81		nat.	70)
LXXV	2	1,98		nat.	

Radical ion of	μ	$a^{\ominus}_{C\mu}$	$a^{\oplus}_{C\mu}$	¹³C*)	Ref.
p-Xylene LXXIX	1	5,20		nat.	188)
	2	6,50		nat.	
Tetracyanoethylene	CN	9,45		nat.	
LXXXVII	1	2,92		nat.	70)
Hexamethyl[3]-radialene XCII	CH_3	4,65		nat.	237)
Hexamethyl[6]radia- -lene XCIII ($R = CH_3$)	7	2,00		nat.	
	CH_3	2,00		nat.	236)
Biacetyl XCIV (trans)	CO	4,5		nat.	288)
Cyclohexane-1,2-dione: XCV (n = 6)	1	4,9		nat.	292)

*) natural abundance; enr. = isotope enriched.
**) The coupling constant strongly dependent on solvent[247].
***)The coupling constant dependent on the nature of the gegenion[248].

The assignment of a $a_{C\mu}$ to a ¹³C nucleus present in one of the n equivalent positions μ is based on the measurement of the intensity of the satellite relative to the intensity of the parent line, on line-width studies[101,153], on selective enrichment with ¹³C at specific sites (see above), and on calculations with the aid of the following equations[110]:

$$a_{C\mu} = Q_C \rho^{\pi}_{\mu} + Q_{C'C}(\rho^{\pi}_{\nu} + \rho^{\pi}_{\nu'}) \tag{26}$$

and

$$a_{C\mu} = Q'_C \rho^{\pi}_{\mu} + Q_{C'C}(\rho^{\pi}_{\nu} + \rho^{\pi}_{\nu'} + \rho^{\pi}_{\nu''}). \tag{27}$$

These equations relate the coupling constant $a_{C\mu}$ of a ¹³C nucleus with the spin population at the centre μ and two or three adjacent carbon centres ν (cf. Section 1.5). The satisfactory agreement between the calculated and the experimental values will be illustrated by the example of the radical-anion $II^{\ominus}$ of naphthalene.

The two coupling constants of ^{13}C nuclei in $II^{\ominus}$ which have been measured by several authors are a_{C1} (positions 1,4,5 or 8) and a_{C2} (positions 2,3,6 or 7). The following values have been obtained from samples either with a natural ^{13}C content [243] or enriched in ^{13}C at one of the above-mentioned positions: [111,242] $a_{C1} = 7.2 \pm 0.1$ and $a_{C2} = 1.1 \pm 0.1$ gauss. Furthermore, a_{C1} has been found to be positive and a_{C2} negative by comparing the line-widths of the ^{13}C satellites in the low-field half and the high-field half of the spectrum[111] (cf. Appendix A.1.3). In accordance with eq. (26), the two ^{13}C coupling constants are calculated from the formulae:

$$a_{C1} = Q_C\,\rho_1^\pi + Q_{C'C}\,(\rho_2^\pi + \rho_9^\pi) \tag{130}$$

$$a_{C2} = Q_C\,\rho_2^\pi + Q_{C'C}\,(\rho_1^\pi + \rho_3^\pi). \tag{131}$$

For the parameters Q_C and $Q_{C'C}$, one can use the values + 35,6 gauss and – 13.9 gauss respectively, proposed by Karplus and Fraenkel[110]. Reliable estimates are also needed for all three spin populations ρ_1^π, ρ_2^π $(= \rho_3^\pi)$ and ρ_9^π, which may be calculated by one of the approximation methods described in Section 1.6. However, more accurate values of ρ_1^π and ρ_2^π at the proton-carrying centres 1 and 2 are obtained from the coupling constants $a_{H1} = 4.95$ and $a_{H2} = 1.83$ gauss with the aid of eq. (20). Use of 24 gauss for the parameter $|Q_{CH}|$ leads to $\rho_1^\pi = 0.206$, and $\rho_2^\pi = 0.076$. The spin population ρ_9^π, which cannot be determined in this manner from experimental data, is found from the condition that the sum of the spin populations ρ_μ^π in a radical-ion must be equal to 1 (eq. (21)). Since theory predicts a positive sign for ρ_1^π and ρ_2^π (cf. Table 2), this condition yields $\rho_9^\pi = -0.069$. By substituting the parameters Q_C and $Q_{C'C}$ and the spin populations ρ_1^π, ρ_2^π $(= \rho_3^\pi)$, and ρ_9^π into eqs. (130) and (131), one obtains the coupling constants a_{C1} +7.24 and $a_{C2} = -1.21$ gauss which are in excellent agreement with the experimental values.

In the case of anthracene (III), all the coupling constants $a_{C\mu}$ (except $a_{C1}^{\oplus}$) have been determined both for the radical-anion and for the radical-cation[153]. As can be seen from Table 13, the experimental values of $a_{C\mu}^{\ominus}$ and $a_{C\mu}^{\oplus}$ for the two corresponding radical-ions $III^{\ominus}$ and $III^{\oplus}$ differ only slightly. This result constitute a further experimental confirmation of the MO statement that, owing to the pairing properties of the system, the spin populations ρ_μ^π in the radical-anion and the radical-cation of a given alternant hydrocarbon should be the same. Recently, two out of three coupling constants $a_{C\mu}$ were also measured for the radical-anion and the radical-cation of biphenylene (X)[327]. The close resemblance of the values $a_{C\mu}^{\ominus}$ and $a_{C\mu}^{\oplus}$ (cf. Table 13) is important in this special case, since the large difference observed for one of the proton splittings ($a_{H1}^{\ominus} = 2.86$; $a_{H1}^{\oplus} = 3.69$ gauss; cf. Table 2) has cast some doubt on the validity of the pairing theorem for the biphenylene system.

A.2.2. Association of radical-anions with alkali metal cations

One speaks of ion-pair formation or association between a radical-anion and a positively charged gegenion when the gegenion persists for a sufficiently long time in the immediate vicinity of the radical-anion, influencing noticeably the properties of the latter. The following discussion will be confined to the best-known case, namely the association of aromatic radical-anions with alkali metal cations.

In the case of such an association, there is an overlap between the MO of the unpaired π-electron in the radical-anion and the orbitals of the gegenion, so that the unpaired electron can delocalize into the gegenion orbitals. The spin population thus donated to the gegenion is generally small. In the case of alkali metal cations as gegenions, however, information from the ESR spectra relevant to this population may become available due to two favourable circumstances: by the s-character of the spin population at the cation, which gives rise to a sufficiently great spin density at the nucleus of the alkali metal cation, and by the magnetic moments of the nuclei, which cause an additional hyperfine splitting. The abundant isotopes, ^{7}Li, ^{23}Na, and ^{39}K, of the most commonly used alkali metalls have a spin quantum number $I = 3/2$ and thus split each hyperfine line of the radical-anion into a set of four equidistant lines of equal intensity (cf. Section 1.1). This type of splitting can be seen in Figure 41, which shows the ESR spectrum of the non-associated radical-anion $XXVIII^{\ominus}$ of 1,4,5,8-tetraazanaphthalene, together with the ESR spectra of the ion-pairs formed between $XXVIII^{\ominus}$ on one hand, and $Na^{\oplus}$ or $K^{\oplus}$, on the other [92]. The associated species may be here considered as complexes, since the lone electron pairs of two nitrogen atoms in peri-positions can participate in bonding to the cation.

Additional hyperfine splitting caused by an alkali metal nucleus is an unambiguous proof of the presence of stable ion-pairs in the solution, i.e. it points to a strong association between the radical-anion and the gegenion. The cation then remains for more than about 10^{-6} sec in the same position relative to the radical-anion. If this residence time is considerably shorter, the association is said to be weak and the ion-pair unstable. In a weak association, the nucleus of the gegenion does not give rise to additional hyperfine splitting. Conversely, however, the absence of such a splitting can not be taken as a sure indication that the association is weak, since, if the spin population in the cation happens to be very small, no hyperfine splitting is observed, although the association may be strong. This applies, e.g. to the associated species in which the gegenion lies in the nodal plane of the singly occupied π-orbital, or in which contributions to the spin population in the cation cancel because of different signs [249]. The influence of weak association was men-

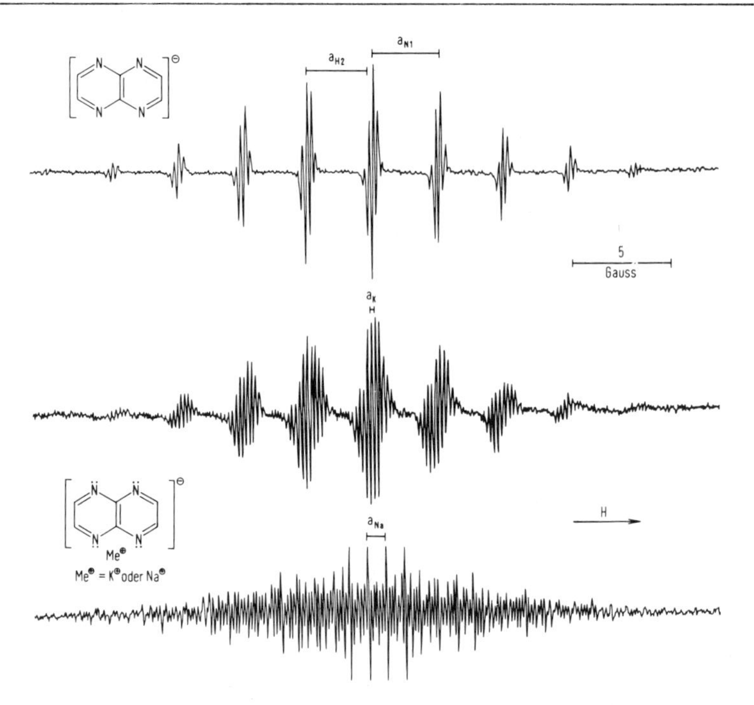

Figure 41: ESR spectra of the radical-anion of 1,4,5,8-tetraazanaphthalene[92].
Top: non-associated species in N,N-dimethylformamide, at +25° C; gegenion: tetraethylammonium⊕;
middle: ion-pair in 1,2-dimethoxyethane, at −50 °C; gegenion: K⊕;
bottom: ion-pair in 1,2-dimethoxyethane, at +25 °C; gegenion: Na⊕.

tioned before in Section 1.3. Although the cation rapidly changes its position relative to the anion, it may cause small fluctuations in the spin populations and thus in the coupling constants of the radical-anion, these fluctuations being then manifested in hyperfine line-broadening. In particularly interesting cases the cation oscillates between two equivalent positions with a frequency that is approximately equal to the induced variations in the coupling constants (expressed in MHz). This leads to anomalous line-widths to be dealt with in Appendix A 2.3.

The associated species formed between the radical-anion of naphthalene (II⊖) and Na⊕ in tetrahydrofuran was the first ion-pair in whose ESR spectra the alkali metal splitting has been extensively investigated[91,250] (shortly before, the phenomenon was observed with the ketyl-anion of benzophenone in 1,2-dimethoxyethane, also with $Na^{\oplus}$ as gegenion[300]). Similar hyperfine splitting has since been reported for the radical-anions of other aromatic hydrocarbons[131,155b,161,251]. The ability of these radical-anions to associate with alkali metal cations in low-polarity solvents such as tetrahydrofuran and 2-methyltetrahydrofuran has also emerged from conductivity measurements and from electronic spectra[252]. As expected, an increase in the dielectric constant of the solvent (use of a more polar solvent or cooling) is accompanied by a weakening of the association. The ability of the solvent to solvate the cations also seems to influence the formation of ion-pairs. Although the dielectric constant of 1,2-dimethoxyethane is smaller than that of tetrahydrofuran[253], the replacement of the latter by the first generally leads to disappearance of an alkali metal splitting in the ESR spectra of hydrocarbon radical-anions associated with their gegenions. This is thought to be due to different solvation of the cations in the two solvents, since 1,2-dimethoxyethane has two oxygen atoms at a distance favourable for complexing the cation[254]. Such complex formation competes with the association between the cation and the radical-anion*).

The dependence of the association on the size of the alkali metal cation is less clear. There are some indications that in the case of hydrocarbon radical-anions, the formation of ion-pairs is favoured by large cations, i.e. that the strength of the association increases along the series : $Li^{\oplus} < Na^{\oplus} < K^{\oplus} < Rb^{\oplus} < Cs^{\oplus}$[153,161,251,252]. This is presumably because solvation of these cations (which competes with association) decreases from left to right in the above series. On the other hand, the reverse order of stability, i.e. $Li^{\oplus} > Na^{\oplus} > K^{\oplus} > Rb^{\oplus} > Cs^{\oplus}$ seems to apply for radical-anions containing heteroatoms with lone pairs[307]. In these cases, the cations probably become attached to the lone pairs, and solvation of these cations will therefore be less important than in the case of pure hydrocarbons. Thus, the ion-pairs, formed by association between alkali metal cations and radical-anions containing nitrogen or oxygen atoms are much tighter than the corresponding associated species which involve hydrocarbon radical-anions.

Relatively stable ion-pairs which exhibit alkali metal splittings even in 1,2-dimethoxyethane have been reported for the radical-anions of many aza-, nitro- and cyano-compounds[92,94,164,165b,172b,255-257,307,313], as well as for some semiquinone-,

---

*) The role of solvation in the formation of ion-pairs has been recently discussed in a number of papers[317-319]. Models involving equilibria between „contact" and „solvent shared" ion-pairs have been suggested[318,319].

semidione- and ketyl-anions[37,248,258,259,287,288,300]. The strongest association is observed in the case of radical-anions possessing two oxygen or nitrogen atoms at a proper distance (2-3 Å) for binding the cation in a manner characteristic of complexes. The radical-anions of 2,2 -bipyridyl (XXXIV) and 1,2-benzoquinone (XLV) surpass even the radical-anion of 1,4,5,8-tetraazanaphthalene in their ability to form such complexes[164,165b,258,260,261]. Beside $XLV^{\ominus}$ other o-semiquinone- and cis-semidione-anions can be expected to give stable complexes of this type.

Table 14 shows the coupling constants a_{Me} of the metal nuclei Me = ^{7}Li, ^{23}Na, and ^{39}K measured for a number of associated species. This list does not include the few reported cases involving the nuclei ^{85}Rb (I = 5/2), ^{87}Rb (I = 3/2), and ^{133}Cs (I = 7/2)[161,249,251].

The coupling constant a_{Me} generally varies with the temperature; cooling of the solution may lead to noticeable increase[37,91] or a decrease[249,251b,256,307,313,314] in a_{Me} (negative or positive temperature coefficient).

Table 14: Coupling constants of the ^{7}Li, ^{23}Na, and ^{39}K nuclei in some alkali metal cations associated with radical-anions

Radical anion of	Me	a_{Me}	Solvent*)	Temp. (°C)**)	Ref.
Naphthalene II	Li	0,19	THF	+23	278)
	Li	0,371	THP	+25	251b)
	Na	1,05	THF	+25	
	Na	1,26	THP	+25	91)
	Na	1,27	DOX	+25	
	Na	1,115	MTHF	+25	
Anthracene III	Na	0,080	THP	+25	251b)
Cyclooctatetraene XII	Li	0,2	THF	0 to + 43	
	Na	0,9	THF	+25	155b)
Biphenyl XIII	Li	0,136	THP	+25	
	Na	0,079	THP	+25	
	K	0,083	THP	+10	251)
	K	0,061	DHP	+10	
	K	0,043	THF	+10	

Radical anion of	Me	a_{Me}	Solvent*)	Temp. (°C)**)	Ref.
Azulene XVI	Li	0,174	THF	+25	
	Na	0,54	THF	+25	262b)
	K	0,2	THF	+25	
Acenaphthylene XVII	Li	0,14	DOX	+23	
	Na	0,06	DOX	+23	
	K	0,07	THF	+23	161)
	K	0,09	DOX	+23	
Pyracene LXV	Na	0,146	THF	-30	
	Na	0,176	MTHF	-80	131)
o-Xylene LXXXI	K	0,17	DME	-80	190)
Pyrazine XXII	Na	0,52	DME	+24	
	Na	0,59	THF	+24	256)
	K	0,10	DME	+24	
1,4,5,8-Tetraaza-	Na	0,95	DME	+25 to –70	92)
naphthalene XXVIII	K	0,20	DME	+25 to –70	
2,2′-Dipyridyl XXXIV	Li	0,72	DOX	+25	164)
1,2-Benzoquinone XLV	Li	0,3	tBA	+25	
	Na	0,2	tBA	+25	258)
Benzophenone XLIX	Li	0,315	DME	+25	279)
	Na	1,13	DME	+20	
	K	0,23	DME	+20	37)
Fluorenone L	Na	0.35	DME	+10	37)
Benzil LII	Na	0,61	THF	-50	259)
m-Dinitrobenzene	Na	0,29	DME	+30 to -40	172b)
LV	K	0,17	DME	+25	280)

Radical anion of	Me	a_{Me}	Solvent*)	Temp.(°C)**)	Ref.
Terephthalo-nitrile LXXV	Na	0,30	DME	+20	
	Na	0,38	THF	+20	
	Na	0,46	MTHF	+20	307)
	K	0,12	THP	+20	
	K	0,13	MTHF	+25	
Phthalonitrile LXXVII	Na	0,30	DME	+20	
	Na	0,28	MTHF	+25	307)
Compound XC***)	K	0,195	DME	+25	97)
Biacetyl XCIV (cis)	Li	0,6	DMSO	+25	288)
Cyclohexane-1,2-dione XCV (n = 6)	Li	0,4	DMSO	+25	287)

*) THF = tetrahydrofuran; MTHF = 2-methyltetrahydrofuran; THP = tetrahydropyran; DHP = dihydropyran; DME = 1,2-dimethoxyethane; DOX = 1,4-dioxane; tBA = t-butyl-alcohol; DMSO = dimethylsulfoxide.

**) +25 °C = room temperature (if no more exact experimental numbers have been given in the reference).

***) formula on page 123.

The spin population at the cation is often calculated with the aid of the equation[251]:

$$a_{Me} = Q_{Me}\,\rho^{ns}_{Me}\,, \qquad (132)$$

where Me = Li, Na, K, etc., n = 2, 3, 4, etc. The values of the hyperfine splittings, which have been measured with atomic beams of the corresponding metals[281], are used for Q_{Me}; in the cases of ^{7}Li, ^{23}Na, and ^{39}K, they are as follows: Q_{Li} = +143, Q_{Na} = +316, and Q_K = +82 gauss.

Applying eq. (132) to the associated species formed between sodium or potassium, on the one hand, and the radical-anion XXVIII$^{\ominus}$ of 1,4,5,8-tetraazanaphthalene on the other, one obtains similar spin populations with a_{Na} = 0.95 and a_K = 0.20 gauss[92]:

$$|\rho^{3s}_{Na}| = 0.0030; \; |\rho^{4s}_{K}| = 0.0024.$$

The simplest quantum-mechanical method yielding finite spin densities at alkali metal nuclei is based on Mulliken's charge transfer (CT) model[282]. In this model the ground configuration $^2\chi_0$, in which the unpaired electron occupies the π-orbital ψ_a of the radical-anion, is mixed with an excited CT configuration $^2\chi_{a,ns}$ resulting from the transfer of this electron into the vacant ns-orbital of the cation:

$$^2\Gamma_0 \approx {}^2\chi_0 + \lambda\, {}^2\chi_{a,ns} \,. \qquad (133)$$

It can be shown that the term of second order in λ in the ground-state function $^2\chi_0$ is mainly responsible for a finite spin density at the alkali metal nucleus. The square $\lambda^2 \ll 1$ corresponds here to the cation's spin population ρ_{Me}^{ns}, which therefore must have a positive sign. Negative contributions to ρ_{Me}^{ns} are possible only when the polarization of the paired electron spins in the cation is taken into account[249]. Since the parameter Q_{Me} is also positive, the coupling constant a_{Me} will have the same sign as the spin population ρ_{Me}^{ns}, i.e. it will generally be positive [cf. eq. (132)].

A.2.3. Time-dependent intramolecular processes

In the last few years some ESR spectra have been found to exhibit an exceptional broadening of certain hyperfine components[131,179,197,262]. The broadening is sometimes so extensive that the affected components escape observation, and the hyperfine splitting seems incompatible with the structure of the radical-ion[53,94,172c,183,191]. This type of line-broadening occurs primarily when the sample exists in two equivalent (or nearly equivalent) forms having the same (or nearly the same) life-time[53,131,179,263,308]:

$$\tau \approx \frac{1}{\gamma_E\,(a_X^I - a_X^{II})}\,, \qquad (134)$$

where γ_E has been defined by eq. (5) of Section 1.1,and a_X^I - a_X^{II} is the difference between the coupling constants of nucleus X in forms I and II. Since this difference lies in the range of 0.1-10 gauss and the ratio γ_E for organic radicals is about 2.8 MHz. gauss ^{1}, the life-time τ varies between 10^{-6} and 10^{-8}sec.
As the simplest example, the hyperfine structure that results from two nuclei A and B, and depends on the life-time τ, will first be considered. If τ is small ($<10^{-8}$ sec). the two nuclei are effectively equivalent. When τ is large ($> 10^{-6}$ sec), on the other hand, the two nuclei give rise to different splittings. In the critical life-time range of 10^{-6} to 10^{-8} sec., anomalous hyperfine structure is observed for nuclei A and B; it will be discussed below in more detail.

When a form I converts into form II or vice versa, nuclei A and B exchange their coupling constants a_A and a_B, but retain their spin quantum numbers M_I (A) and M_I (B), i.e.

$$a_A^I = a_B^{II} \text{ and } a_A^{II} = a_B^I; \quad \text{but} \quad M_I^I(A) = M_I^{II}(A) = M_I(A) \quad \text{and} \quad M_I^I(B) = M_I^{II}(B) = M_I(B)$$

The position of a hyperfine line in the ESR spectra of forms I and II can therefore be specified as follows:

$$\begin{aligned} &\text{form I: } H^I = H_O + a_A M_I(A) + a_B M_I(B); \\ &\text{form II: } H^{II} = H_O + a_B M_I(A) + a_A M_I(B), \end{aligned} \tag{136}$$

where H^I and H^{II} denote the field strengths at which the hyperfine line for forms I and II is observed; H_o is the field at the centre of the spectrum; $M_I(A)$ and M_I (B) are those spin quantum numbers of nuclei A and B which have been assigned to the line (cf. Appendix A.1.3).

The contribution $(\Delta\nu)_{exch}$ to the line-width $\Delta\nu$ (= $\gamma_E \Delta H$) coming from the interconversions I → II and II → I, depends both on the life-time τ of a form and on the square of the shift (H^I- H^{II}), i.e.

$$(\Delta\nu)_{exch} = \tfrac{\tau}{8}\gamma_E^2(H^I - H^{II})^2 \tag{137}$$

This shift is derived from eqs. (136) as

$$H^I - H^{II} = (a_A - a_B)[M_I(A) - M_I(B)] \tag{138}$$

so that

$$(\Delta\nu)_{exch} = [\tfrac{\tau}{8}\gamma_E^2(a_A - a_B)^2][M_I(A) - M_I(B)]^2. \tag{139}$$

When A and B belong to two equally large sets of n equivalent nuclei A_i and B_i, the spin quantum numbers $M_I(A)$ and $M_I(B)$ are to be replaced by the sums $\sum_{i=1}^{n} M_I(A_i)$ and $\sum_{i=1}^{n} M_I(B_i)$, and thus

$$(\Delta\nu)_{exch} = [\tfrac{\tau}{8}\gamma_E^2(a_A - a_B)^2]\left[\sum_{i=1}^{n} M_I(A_i) - \sum_{i=1}^{n} M_I(B_i)\right]^2. \tag{140}$$

For an ESR spectrum recorded under fixed conditions, the first square-bracketted term in eqs. (139) and (140) has a constant value, and the variation in $(\Delta\nu)_{exch}$ is therefore determined by the square of $[M_I(A)-M_I(B)]$ or $[\sum_i M_I(A_i) - \sum_i M_I(B_i)]$.

This is shown in Figure 42 by two schemes, which refer to two pairs of equivalent protons A_i and B_i, the summed spin quantum numbers $\sum_{i=1}^{2} M_I(A_i)$ and $\sum_{i=1}^{2} M_I(B_i)$ having thus the values −1,0 and +1 (cf. Figure 2)

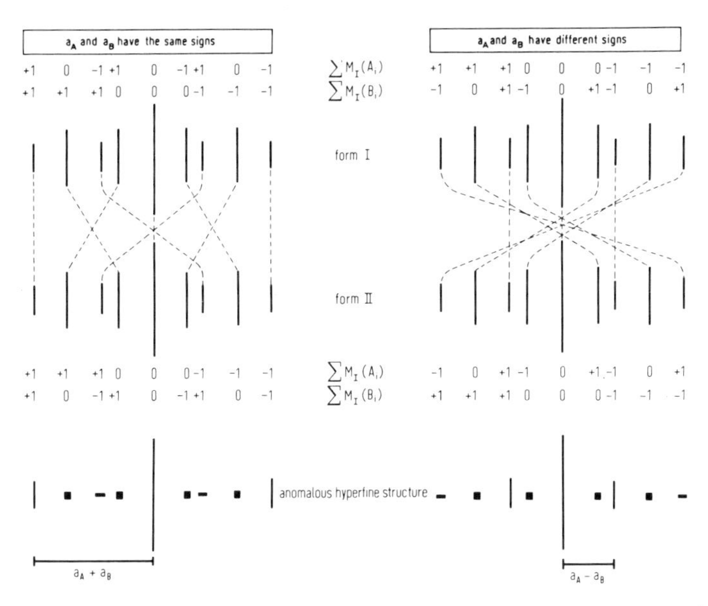

Figure 42: Line-broadening resulting from exchange between forms I and II in the case of two pairs of equivalent protons (see text).

It can be seen from Figure 42 that the hyperfine lines undergoing the greatest shift by interconversions of forms I and II are those which exhibit the largest differences

$$\left|\sum_{i=1}^{2} M_I(A_i) - \sum_{i=1}^{2} M_I(B_i)\right)$$

In accordance with eq. (140), these lines are broadened the most. Only the three lines whose position is unchanged in the spectra of forms I and II, and for which $\sum_{i=1}^{2} M_I(A_i) = \sum_{i=1}^{2} M_I(B_i)$, remain narrow. They can always be observed in the ESR spectrum, whereas the other components often broaden beyond recognition. The relative intensities of the three lines are 1:4:1; their distance is $|a_A| \pm |a_B|$, according to whether the two coupling constants have the same or different signs. The first case (a_A and a_B of the same sign) is illustrated by the spectra of 1,2,3-trihydropyrenyl[265] reproduced in Figure 43.

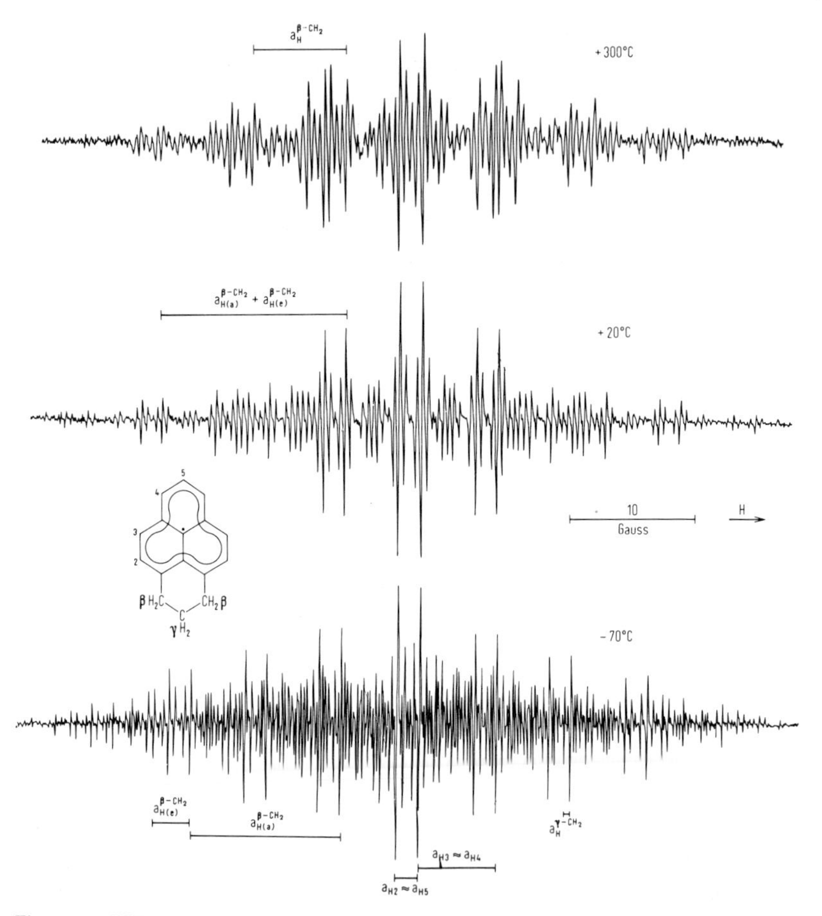

Figure 43: ESR spectra of 1,2,3-trihydropyrenyl.
Top: in 1-bromonaphthalene, at +300 °C;
middle: in 1-bromonaphthalene at +25 °C;
bottom: in 1,2-dimethoxyethane, at −70 °C.

In the trimethylene chain of this neutral radical, interconversion occurs between the axial and the equatorial positions of the β-protons. The life-time of a conformation at room temperature satisfies condition (134), and the β-protons give rise to three groups of lines with an intensity distribution of 1:4:1 (cf. Figure 43 middle). The distance between the groups is equal to the sum of the coupling constants of an

axial (a) and an equatorial (e) proton, i.e. $a_{H(a)}^{\beta\text{-}CH_2} + a_{H(e)}^{\beta\text{-}CH_2}$.
When the solution is cooled to -70 °C, the life-time τ becomes sufficiently long to permit the measurement of the individual β-proton coupling constants. One thus obtains a "normal" spectrum, which contains three times as many lines as that recorded at room temperature (cf. Figure 43, bottom). On the other hand, strong heating of the solution also leads to a drastic change in the appearance of the spectrum. At $+300$ °C, the life-time τ is so short that the two pairs of β-protons are magnetically equivalent (cf. Figure 43, top), and their coupling constant is then equal to

$$a_{H}^{\beta-CH_2} = \tfrac{1}{2}\,(a_{H(a)}^{\beta-CH_2} + a_{H(e)}^{\beta-CH_2}).$$

Similar considerations also explain anomalous line-widths in many other cases. The following time-dependent intramolecular processes are responsible for the line-broadening:

1. Conformational interconversions of a methylene chain, as in 1,2,3-trihydropyrenyl. They are effective in the radical-ions of the tetrahydro- and the hexahydro-derivatives of pyrene (CIV and CV)[53,191].

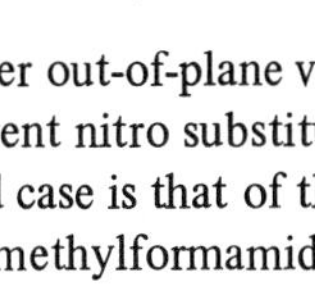

CIV 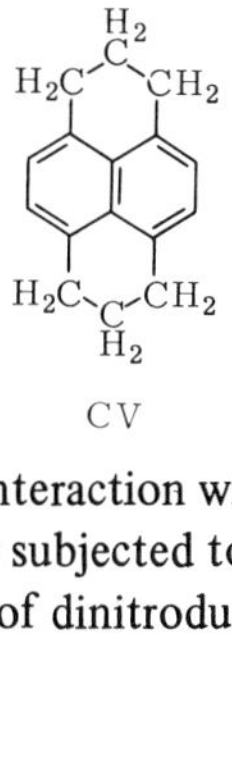 CV

2. Either out-of-plane vibrations or an interaction with the solvent of two formally equivalent nitro substituents alternately subjected to perturbation. An extensively studied case is that of the radical-anion of dinitrodurene(CVI)[179a,c] dissolved in N,N-dimethylformamide.

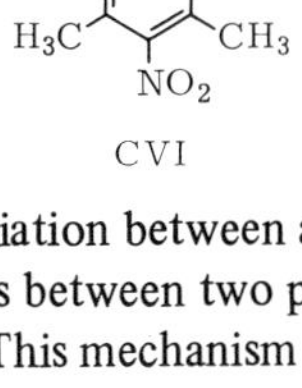

CVI

3. Association between a radical-anion and an alkali metal cation, when the cation oscillates between two positions equivalent with respect to the anion (cf. Appendix A 2.2). This mechanism is exemplified by the associated species formed by the

radical-anions of pyrazine (XXII)[255b,256], pyracene (LXV)[131], terephthalonitrile (LXXV)[307,313] and m-dinitrobenzene (LV)[315].

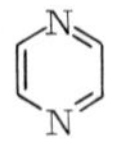

XXII 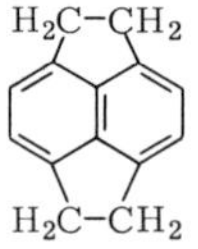 LXV 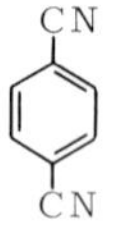 XXV 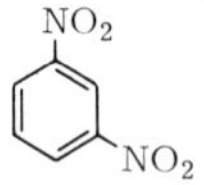 LV

4. Cis-trans isomerism due to restricted rotation about a formally single bond, as is encountered in the radical-cations of dihydroduroquinone (CVII)[197] and 1,4,5,8-tetrahydroxynaphthalene (CVIII)[266].

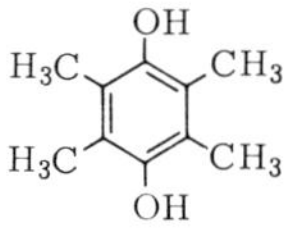

CVII 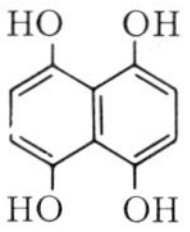 CVIII

Anomalous line-widths may also arise as a result of intramolecular processes involving not the nuclei of the radical-ions, but the unpaired electrons. It was mentioned in Section 1.3 that the electron-exchange between a radical-anion on one hand, and either the corresponding unreduced compound or the diamagnetic dianion on the other, can influence the line-width of the ESR spectra of radical-anions. Besides an intermolecular exchange of this type, there may also occur an intramolecular electron transfer, whenever two or more equivalent (or nearly equivalent) aromatic residues are not too widely separated by a saturated group. The hyperfine structure of the radical-ion then depends on the time τ for which the unpaired electron stays with one residue. If τ is longer than ca. 10^{-6} sec, the splitting is determined only by the nuclei in the residue in question. If τ is shorter than ca. 10^{-8} sec, the hyperfine structure is due to the nuclei in both residues. Again anomalous line-widths occur in an intermediate range where the transfer frequency $1/\tau$ is of the order of magnitude of the coupling constant expressed in MHz [cf. eq. (134)]. Relevant examples are the radical-anions of [m,n]paracyclophanes (CIX)[267] and the bis-p-nitrophenyl compounds (CX)[268].

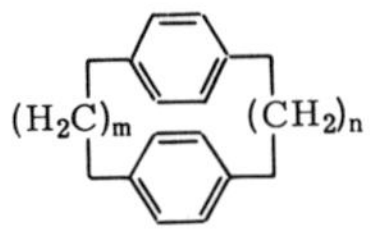

CIX 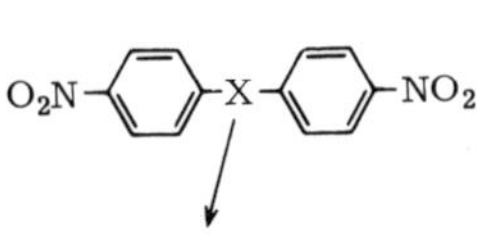 CX $X = O, S, CH_2$

The theory of anomalous line-widths resulting from time-dependent intramolecular processes has been developed by various authors[269,179c,270,]8)]. The term 'alternating' line-widths is often used in the literature[131,179,191,263]. In fact, an alternation of narrow and broad hyperfine components (or groups of components) can often be observed in ESR spectra exhibiting the anomalous line-widths which have been considered in this Appendix.

Bibliography

1) A. Streitwieser Jr.: Molecular Orbital Theory for Organic Chemists. Wiley, New York 1961, Chapter 6.
2) A. Carrington, Quart. Reviews *17,* 67 (1963).
3) J.E. Wertz, Chem. Reviews *55,* 829 (1955).
4) D.H. Whiffen, Quart. Reviews *12,* 250 (1958).
5) D.J.E. Ingram: Free Radicals as studied by Electron Spin Resonance. Butterworth, London 1958.
6) H.A. Staab: Einführung in die theoretische organische Chemie. Verlag Chemie, Weinheim 1959, Chapter 2,8.
7) The NMR-ESR Staff of Varian Ass.: NMR and ESR Spectroscopy. Pergamon Press, London 1960.
8) W. Low: Paramagnetic Resonance in Solids. Solid State Physics, Suppl. 2, Academic Press, New York 1960.
9) K.H. Hausser: Magnetische Elektronen- und Kernresonanz, in: Ullmanns Encyklopädie der technischen Chemie. 3rd edit., Urban und Schwarzenberg, München 1961, Vol. II/1.
10) G.E. Pake: Paramagnetic Resonance. Benjamin, New York 1962.
11) Ch.P. Slichter: Principles of Magnetic Resonance. Harper and Row, New York 1963.
12) M.C.R. Symons: The Identification of Organic Free Radicals by Electron Spin Resonance, in V. Gold: Advances in Physical Organic Chemistry. Academic Press, New York 1963, Vol. I.
13) A. Horsfield, Chimia *17,* 42 (1963).
14) G. Schoffa: Elektronenspinresonanz in der Biologie. Springer, Heidelberg 1964.
15) A.L. Buchachenko: Stable Radicals. Consultants Bureau, New York 1965.
16) H. Fischer: Magnetische Eigenschaften freier Radikale, in Landolt-Börnstein: Zahlenwerte und Funktionen. New series, Group II, Vol. I. Springer, Heidelberg 1965.
17) G. Herzberg: Atomic Spectra and Atomic Structure. 2nd edit., Dover Publ., New York 1944.
18) J.A. Pople, W.G. Schneider, and H.J. Bernstein: High Resolution Nuclear Magnetic Resonance. McGraw-Hill, New York 1959.
19) E. Fermi, Z. Physik *60,* 320 (1930).
20) M. Gomberg, Ber. dtsch. chem. Ges. *33*, 3150 (1900); J.Amer. chem. Soc. *22,* 757 (1900).
21) M. Gomberg and W.E. Bachmann, J. Amer. chem. Soc. *49,* 236 (1927); W.E. Bachmann, ibid. *55,* 1179 (1933); R.N. Doescher and G.W. Wheland, ibid. *56,* 2011 (1934); S. Sudgen, Trans.Faraday Soc. *30,* 18 (1934).

22) W. Schlenk and E. Bergmann, Liebigs Ann.Chem. *463,* 1 (1928); N.D. Scott, J.F. Walker, and V.L. Hansley, J.Amer.chem.Soc. *58,* 2442 (1936).
23) L. Michaelis, J. biol. Chemistry *92,* 211 (1931); L. Michaelis, M.P. Schubert, R.K. Reber, J.A. Kuck, and S. Granick, J.Amer.chem. Soc. *60,* 1678 (1938); L. Michaelis and S. Granick, ibid. *70,* 624 (1948); L. Michaelis, Ann.N.Y. Acad.Sci. *40,* 39 (1940).
24) a) S.I. Weissman, J. Townsend, D.E. Paul, and G.E. Pake, J.chem.Physics *21,* 2227 (1953); b) T.L. Chu, G.E. Pake, D.E. Paul, J. Townsend, and S.I. Weissman, J.physic.Chem. *57,* 504 (1953); c) D. Lipkin, D.E. Paul, J. Townsend, and S.I. Weissman, Science *117,* 534 (1953).
25) B. Venkataraman and G.K. Fraenkel, a) J.Amer.chem.Soc. *77,* 2707 (1955); b) J.chem.Physics *23,* 588 (1955).
26) N.S. Hush and J.A. Pople, Trans.Faraday Soc. *51,* 600 (1955).
27) M.E. Wacks and V.H. Dibeler, J.chem.Physics *31,* 1557 (1959).
28) A. Streitwieser Jr.: Molecular Orbital Theory for Organic Chemists. Wiley, New York 1961.
29) E. de Boer and S.I. Weissman, J.Amer.chem.Soc. *80,* 4549 (1958).
30) T.R. Tuttle and S.I. Weissman, J.Amer.chem.Soc. *80,* 5342 (1958).
31) C.A. McDowell, K.F. Paulus, and J.R. Rowlands, Proc.chem.Soc. (London) *1962,* 60.
32) A. Carrington and J. dos Santos-Veiga, Mol.Physics *5,* 21 (1962).
33) R.L. Ward, J.Amer.chem.Soc. *84,* 332 (1962).
34) G.J. Hoijtink, E. de Boer, P.H. van der Meij, and W.P. Weijland, Recueil Trav. chim. Pays-Bas, *74,* 277 (1955); *75,* 487 (1956); P. Balk, G.J. Hoijtink, and J.W.H. Schreuers, ibid. *76,* 813 (1957); E. de Boer and S.I. Weissman, ibid. *76,* 824 (1957); J. Hoijtink and P.H. van der Meij, Z.physik.Chem. (Frankfurt/M) *20,* 1 (1959).
35) N.S. Hush and J.R. Rowlands, J.chem.Physics *25,* 1076 (1956).
36) J.R. Bolton, Mol. Physics *6,* 219 (1963).
37) P.B. Ayscough and R. Wilson, Proc.chem.Soc. (London) *1962,* 229; J.chem.Soc. (London) *1963,* 5412.
38) P.B. Ayscough, F.B. Sargent, and R. Wilson, J.chem.Soc. (London) *1963,* 5418.
39) P.L. Kolker and W.A. Waters, Proc.chem.Soc. (London) *1963,* 55; J.chem.Soc. (London) *1964,* 1136.
40) A. Maximadshy and F. Dörr, Z.Naturforsch. *19b,* 359 (1964); E. Brunner and F. Dörr, Ber.Bunsenges.physik.Chem. *68,* 468 (1964).
41) V. Yokozawa and I. Miyashita, J.chem.Physics *25,* 796 (1956).
42) G.J. Hoijtink and W.P. Weijland, Recueil Trav.chim.Pays-Bas *76,* 836 (1957).
43) S.I. Weissman, E. de Boer, and J.J. Conradi, J.chem.Physics *26,* 963 (1957).

44) A. Carrington, F. Dravnieks, and M.C.R. Symons, J.chem.Soc. (London) *1959,* 947.
45) J.R. Bolton, quoted in 59).
46) J.R. Bolton and A. Carrington, Mol.Physics *4,* 271 (1961).
47) C.A. McDowell and J.R. Rowlands, Canad.J.Chem. *38,* 503 (1960).
48) A. Carrington and J. dos Santos-Veiga, Mol.Physics *5,* 285 (1962).
49) F. Gerson and J. Heinzer, Chem.Commun. (London) *1965,* 488; Helv.chim.Acta *49,* 7 (1966).
50) J.A. Brivati, R. Hulme, and M.C.R. Symons, Proc.chem.Soc. (London) *1961,* 384.
51) J.R. Bolton, A. Carrington, and A.D. McLachlan, Mol.Physics *5,* 31 (1962).
52) E. de Boer and E.L. Mackor, Mol.Physics *5,* 493 (1962).
53) E. de Boer and A.D. Praat, Mol.Physics *8,* 291 (1964).
54) F. Gerson, E. Heilbronner, and V. Boekelheide, Helv.chim.Acta *47,* 1123 (1964).
55) F. Gerson and J. Heinzer, Helv.chim.Acta *50,* 1852 (1967).
56) a) J.E. Wertz and J.L. Vivo, J.chem.Physics *23,* 2193 (1955); A. Fava, P.B. Sogo, and M. Calvin, J.Amer.chem. Soc. *79,* 1078 (1957); b) H.J. Shine, C.F. Dais, and R.J. Small, J. org. Chemistry *29,* 21 (1964).
57) E.A.C. Lucken, a) J. chem.Soc. (London) *1962,* 4963; b) Theor.chim.Acta *1,* 397 (1963).
58) F. Gerson, Helv.chim.Acta *47,* 1484 (1964).
59) I.C. Lewis and L.S. Singer, J.chem.Physics *43,* 2712 (1965); *44,* 2082 (1965).
60) a) J.R. Bolton and A. Carrington, Proc.chem.Soc. (London) *1961,* 385; b) J.R. Bolton, A. Carrington, and J. dos Santos-Veiga, Mol.Physics *5,* 465 (1962).
61) S. Wawzonek and H.A. Laitinen, J.Amer.chem.Soc. *64,* 2365 (1942); S. Wawzonek and J. Wang-Fan, ibid. *68,* 2541 (1946).
62) G.J. Hoijtink and J. van Schooten, Recueil Trav.chim.Pays-Bas *71,* 1089 (1952); *72,* 691 (1953); G.J. Hoijtink, J. van Schooten, E. de Boer, and W.Ij. Aalbersberg, ibid. *73,* 355 (1954); G.J. Hoijtink, ibid. *73,* 895 (1954).
63) H. Lund, Acta chem. Scand. *11,* 1323 (1957).
64) D.H. Geske and A.H. Maki, J.Amer.chem.Soc. *82,* 2671 (1960).
65) M.T. Melchior and A.H. Maki, J.chem.Physics *34,* 471 (1961).
66) K.H. Hauser, A. Häbich, and V. Franzen, Z.Naturforsch. *16a,* 836 (1961).
67) I. Bernal, P.H. Rieger, and G.K. Fraenkel, J.chem.Physics *37,* 1489 (1962).
68) P.H. Rieger and G.K. Fraenkel, J.chem.Physics *37,* 2795 (1962).
69) P.H. Rieger and G.K. Fraenkel, J.chem.Physics *37,* 2811 (1962).
70) P.H. Rieger, I. Bernal, W.H. Reinmuth, and G.K. Fraenkel, J.Amer.chem.Soc. *85,* 683 (1963).
71) R. Dehl and G.K. Fraenkel, J.chem.Physics *39,* 1793 (1963).
72) B.L. Barton and G.K. Fraenkel, J.chem.Physics *41,* 1455 (1964).
73) D.H. Levy and R.J. Myers, J.chem.Physics *41,* 1062 (1964).

74) D.H. Levy, Ph.D. Thesis, University of California, Berkeley (1965).
75) J.M. Fritsch and R.N. Adams, J.chem.Physics *43,* 1887 (1965).
76) K. Kuwata and D.H. Geske, J.Amer.chem.Soc. *86,* 2101 (1964).
77) P. Smejtek, J. Honzl, and V. Metalova, Coll.Czechoslov.chem.Commun. *30,* 3875 (1965).
78) B.G. Segal, M. Kaplan, and G.K. Fraenkel, J.chem.Physics *43,* 4191 (1965).
79) G. Henrici-Olivé and S. Olivé, Z.physik.Chem. (Frankfurt/M) *42,* 145 (1964); *43,* 327 (1964).
80) R. Hulme and M.C.R. Symons a) Proc.chem.Soc. (London) *1963,* 241; b) J.chem.Soc. (London) *1965,* 1120.
81) K.H. Hausser, Proc. Xth Colloquium Internationale, College Park, 1962, pp. 708-718.
82) F. Gerson, B. Weidmann, and E. Heilbronner, Helv.chim.Acta *47,* 1951 (1964).
83) G.E. Pake and T.R. Tuttle, Physic.Review Letters *3,* 423 (1959).
84) K.H. Hausser, Naturwissenschaften *47,* 251 (1960).
85) R.L. Ward and S.I. Weissman, J.Amer.chem.Soc. *76,* 3612 (1954).
86) M.T. Jones and S.I. Weissman, J.Amer.chem.Soc. *84,* 4269 (1962).
87) P. Ludwig and R.N. Adams, J.chem.Physics *37,* 828 (1962).
88) a) N.M. Atherton, F. Gerson, and J.N. Murrell, Mol.Physics *6,* 265 (1963); b) F. Gerson and J.D.W. van Voorst, Helv.chim.Acta *46,* 2257 (1963).
89) Handbook of Chemistry and Physics, 44th edit., The.Chem.Rubber Publ. Co., Cleveland, Ohio 1962, p. 2263.
90) J.S. Hyde and H.W. Brown, J.chem.Physics *37,* 368 (1962).
91) N.M. Atherton and S.I. Weissman, J.Amer.chem.Soc. *83,* 1330 (1961).
92) F. Gerson and W.F.L. Armarego, Helv.chim.Acta *48,* 112 (1965).
93) S.H. Glarum and L.C. Snyder, J.chem.Physics *36,* 2989 (1962).
94) F. Gerson and R.N. Adams, Helv.chim.Acta *48,* 1539 (1965).
95) T.R. Tuttle, R.L. Ward, and S.I. Weissman, J.chem.Physics *25,* 189 (1956).
96) S.I. Weissman, T.R. Tuttle, and E. de Boer, J.physic. Chem. *61,* 28 (1957).
97) F. Gerson, E. Heilbronner, W.A. Böll, and E. Vogel, Helv.chim.Acta *48,* 1494 (1965).
98) A.I. Schatenstein: Isotopenaustausch und Substitution des Wasserstoffs in organischen Verbindungen. Deutscher Verlag der Wissenschaften, Berlin 1963, Anhang 6.3.
99) G.T. Jones and J.N. Murrell, Chem.Commun. (London) *1965,* 28.
100) J.C. Schug, T.H. Brown, and M. Karplus, J.chem.Physics *37,* 330 (1962).
101) J.R. Bolton and G.K. Fraenkel, J.chem.Physics *41,* 944 (1964).
102) G.E. Pake, J. Townsend, and S.I. Weissman, Physic. Reviews *85,* 682 (1952); G.E. Pake, S.I. Weissman, and J. Townsend, Disc.Faraday Soc. *19,* 147 (1965).

103) H.M. McConnell, J.chem.Physics *24,* 764 (1956).
104) S.I. Weissman, J.chem.Physics *25,* 890 (1956).
105) R. Bersohn, J.chem.Physics *24,* 1066 (1956).
106) H.S. Jarrett, J.chem.Physics *25,* 1289 (1956).
107) H.M. McConnell and D.B. Chesnut, J.chem.Physics *28,* 107 (1958).
108) H.M. McConnell, J.chem.Physics *24,* 632 (1956).
109) H.M. McConnell and D.B. Chesnut, J.chem.Physics *27,* 984 (1957).
110) M. Karplus and G.K. Fraenkel, J.chem.Physics *35,* 1312 (1961).
111) E. de Boer and E.L. Mackor, J.chem.Physics *38,* 1450 (1963).
112) N.M. Atherton, F. Gerson, and J.N. Murrell, Mol.Physics *5,* 509 (1962).
113) E.W. Stone and A.H. Maki, J.chem.Physics *39,* 1635 (1963).
114) C.A. McDowell and K.F.G. Paulus, Mol.Physics *7,* 541 (1963-64).
115) G. Favini and A. Gamba, Ricerca Sci. *6,* A, 383 (1964).
116) J.C.M. Henning, J.chem.Physics *44,* 2139 (1966).
117) B.L. Barton and G.K. Fraenkel, J.chem.Physics *41,* 695 (1964).
118) D. Pooley and D.H. Whiffen, Mol.Physics *4,* 81 (1961).
119) R.W. Fessenden and R.H. Schuler, J.chem.Physics *39,* 2147 (1963).
120) N.M. Atherton, E.J. Land, and G. Porter, Trans.Faraday Soc. *59,* 818 (1963).
121) A.D. McLachlan, Mol.Physics *1,* 233 (1958).
122) D.B. Chesnut, J.chem.Physics *29,* 43 (1958).
123) J.P. Colpa and E. de Boer, Physics Letters *5,* 225 (1963); Mol. Physics *7,* 333 (1963-64).
124) D.H. Levy, Mol.Physics *10,* 233 (1966).
125) R.S. Mulliken, J.chem.Physics *7,* 339 (1939); R.S. Mulliken, C.A. Rieke, and W.G. Brown, J.Amer.chem.Soc. *63,* 41 (1941).
126) C.A. Coulson and V.A. Crawford, J.chem.Soc. (London) *1953,* 2052.
127) P.G. Lykos, J.chem.Physics *32,* 625 (1960).
128) C. Heller and H.M. McConnell, J.chem.Physics *32,* 1535 (1960).
129) A. Horsfield, J.R. Morton, and D.H. Whiffen, Mol.Physics *4,* 425 (1961).
130) F. Gerson, J. Heinzer, and B. Weidmann, unpublished work.
131) E. de Boer and E.L. Mackor, Proc.chem.Soc. (London) *1963,* 23; J.Amer.chem.Soc. *86,* 1513 (1964).
132) C. de Waard and J.C.M. Henning, Physics Letters *4,* 31 (1963).
133) E. Hückel, Z.Physik. *70,* 204 (1931); *72,* 310 (1931); *76,* 628 (1932); *86,* 632 (1933).
134) C.A. Coulson: Valence, 2nd edit., Oxford University Press 1961, Chapter 9.6.
135) R. Daudel, R. Lefebvre, and C. Moser: Quantum Chemistry, 2nd edit. Interscience, New York 1961.
136) J.N. Murrell, S.F.A. Kettle, and M. Tedder: Valence Theory. Wiley, New York 1965.

137) C.A. Coulson and H.C. Longuet-Higgins, Proc.Roy.Soc. *A 192,* 16 (1947).
138) G. Vincow and G.K. Fraenkel, J. chem.Physics *34*, 1333 (1961).
139) P. Brovetto and S. Ferroni, Nuovo Cimento *5*, 142 (1957).
140) D.R. Hartree, Proc.Cambridge Phil.Soc. *24*, 89,111 (1928); V. Fock, Z.Physik. *61,* 126 (1930).
141) C.C.J. Roothaan, Reviews mod. Physics, *23,* 69 (1951).
142) J.A. Pople, Trans.Faraday Soc. *49*, 1375 (1953).
143) J.A. Pople and R.K. Nesbet, J.chem.Physics *22*, 571 (1954).
144) T. Amos and L.C. Snyder, J.chem.Physics *41,* 1773 (1964); L.C. Snyder and T. Amos, ibid. *42,*3670 (1965).
145) A.D. McLachlan, Mol.Physics *3*, 233 (1960).
146) A.D. McLachlan, Mol.Physics *2*, 271 (1959); ibid. *4*, 49 (1961).
147) G.J. Hoijtink, Mol.Physics *1*, 157 (1958).
148) R. Pariser and R.G. Parr, J.chem.Physics *21,* 466, 767 (1953); R. Pariser, ibid. *21,* 568 (1953); *24,* 250 (1956); *25,* 1112 (1956).
149) G.J. Hoijtink, J. Townsend, and S.I. Weissman, J.chem.Physics *34,* 507 (1961).
150) M.E. Anderson, P.J. Zandstra, and T.R. Tuttle, J.chem.Physics *33,* 1591 (1960).
151) P.B. Sogo, M. Nakazaki, and M. Calvin, J.chem.Physics *26,* 1343 (1957).
152) H.H. Dearman and H.M. McConnell, J.chem.Physics *28,* 51 (1958).
153) J.R. Bolton and G.K. Fraenkel, J.chem.Physics *40*, 3307 (1964).
154) J.P. Colpa and J.R. Bolton, Mol.Physics *6*, 273 (1963).
155) a) T.J. Katz and H.L. Strauss. J.chem.Physics *32*, 1873 (1960);
b) H.L. Strauss, T.J. Katz, and G.K. Fraenkel, J. Amer.chem.Soc. *85*, 2360 (1963).
156) K.H. Hausser, L. Mongini, and R. van Steenwinkel, Z.Naturforsch. *19a*, 777 (1964
157) N.M. Atherton, F. Gerson, and J.N. Ockwell, J.chem.Soc. (London) *A 1966,* 109.
158) J.R. Bolton, J.chem.Physics *43,* 309 (1965).
159) G.Giacometti, P.L. Nordio, and M.V. Pavan, Theor.chim.Acta *1*, 404 (1963).
160) EUCHEM Conference on Chemical Aspects of Electron Spin Resonance, Cirencester, England 1965.
161) M. Iwaizumi and I. Isobe, Bull.chem.Soc. Japan, *37,* 1651 (1964).
162) F. Gerson and B. Weidmann, Helv.chim.Acta *49,* 1837 (1966).
163) K. Markau and W. Maier, Z.Naturforsch. *16a,* 1116 (1961).
164) E. König and H. Fischer, Z.Naturforsch. *17a,* 1063 (1962).
165) a) F. Bruin, F.W. Heineken, M. Bruin, and A. Zahlan, J.chem.Physics *36,* 2783 (1962); b) A. Zahlan, F.W. Heineken, M. Bruin, and F. Bruin, ibid. *37,* 683 (1962)
166) F. Gerson and G. Wolf, unpublished work.
167) R. L. Myers and C.L. Talcott, Mol. Physics *12,* 549 (1967).
168) E.W. Stone and A.H. Maki, J.chem.Physics a) *36,* 1944 (1962); b) *41,* 284 (1964).

169) N. Steinberger and G.K. Fraenkel, J.chem. Physics *40,* 723 (1964).
170) E.J. Land and G. Porter, Proc.chem.Soc. London *1960,* 84.
171) R.L. Ward and M.P. Klein, J.chem.Physics *28,* 518 (1958); *29,* 678 (1958); R.L. Ward, ibid, *30,* 852 (1959).
172) R.L. Ward, a) J.chem.Physics *32,* 410 (1960); b) ibid. *36,* 1405 (1962); c) J.Amer.chem.Soc. *83,* 1296 (1961).
173) A.H. Maki and D.H. Geske, J.chem.Physics *33,* 825 (1960).
174) a) J.Q. Chambers, T. Layloff, and R.N. Adams, J.physic.Chem. *68,* 661 (1964); b) P. Ludwig, T. Layloff, and R.N. Adams, J. Amer.chem.Soc. *86,* 4568 (1964): c) T. Kitagawa, T. Layloff, and R.N. Adams, Analytic.Chem. *36,* 925 (1964).
175) J. Pannell, Mol.Physics *7,* a) 317; b) 599 (1963-64).
176) P.H. Rieger and G.K. Fraenkel. J.chem.Physics *39,* 609 (1963).
177) T. Fujinaga, Y. Deguchi, and K. Vemoto, Bull.chem.Soc. Japan *37,* 822 (1964).
178) M.J. Blandamer, T.E. Gough, J.M. Gross, and M.C.R. Symons, J.chem.Soc. (London) *1964,* 536.
179) a) J.H. Freed and G.K. Fraenkel, J.chem.Physics *37,* 1156 (1962); b) J.H. Freed, P.H. Rieger, and G.K. Fraenkel, ibid. *37,* 1881 (1962); c) J.H. Freed and G.K. Fraenkel, ibid. *41,* 699 (1964).
180) S.H. Glarum and J.H. Marshall, J.chem.Physics *41,* 2182 (1964).
181) P.H.H. Fischer and C.A. McDowell, Mol.Physics *8,* 357 (1964).
182) D.H. Levy and R.J. Myers, J.chem.Physics *42,* 3731 (1965).
183) J.M. Gross and M.C.R. Symons, Mol.Physics *9,* 287 (1965).
184) P.H.H. Fischer and C.A. McDowell, Canad.J.Chem. *43,* 3400 (1965).
185) J.M. Gross and M.C.R. Symons, J.chem.Soc. (London) *A 1966,* 451.
186) K.H. Hausser, a) Z.Naturforsch. *14a,* 425 (1959); b) Mol.Physics *7,* 195 (1963).
187) T.R. Tuttle, J.Amer.chem.Soc. *84,* 2839 (1962).
188) J.R. Bolton and A. Carrington, Mol.Physics *4,* 497 (1961).
189) J.R. Bolton, A. Carrington, A. Forman, and L.E. Orgel, Mol.Physics *5,* 43 (1962).
190) J.R. Bolton, J.chem.Physics *41,* 2455 (1964).
191) M. Iwaizumi and T. Isobe, Bull.chem.Soc. Japan *38,* 1547 (1965).
192) J. dos Santos-Veiga, Rev.Port.Quim. *6,* 1 (1964).
193) A. Carrington and P.F. Todd, a) Mol.Physics *7,* 533 (1963-64); b) ibid. *8,* 299 (1964).
194) K. Ishizu, Bull.chem.Soc. Japan *36,* 939 (1963); *37,* 1093 (1964).
195) J.D. Roberts: Notes on Molecular Orbital Calculations. Benjamin, New York 1962.
196) a) B. Venkataraman, B.G. Segal, and G.K. Fraenkel, J.chem.Physics *30,* 1006 (1959); b) G.K. Fraenkel, Ann.N.Y.Acad.Sci. *67,* 546 (1957).
197) J.R. Bolton and A. Carrington, Mol.Physics *5,* 161 (1962).

198) K. Maruyama, R. Tanikaga, and R. Goto, J.chem.Soc Japan *84,* 75 (1963); Bull.chem.Soc. Japan *36,* 1141 (1963).
199) A.H. Maki and D.H. Geske, J.Amer.chem.Soc. *83,* 1852 (1961).
200) Z. Galus and R.N. Adams, J.chem.Physics *36,* 2814 *(1962).*
201) J.E. Wertz and J.L. Vivo, J.chem.Physics *23,* 2441 (1956).
202) A. Carrington, A. Hudson, and H.C. Longuet-Higgins, Mol.Physics *9,* 377 (1965).
203) R.L. Ward, J.chem.Physics *32,* 1592 (1960).
204) A. Carrington and P.F. Todd, Mol.Physics *6,* 161 (1963).
205) M.C. Townsend and S.I. Weissman, J.chem.Physics *32,* 309 (1960).
206) H.A. Jahn and E. Teller, Proc.Roy.Soc. *A 161,* 220 (1937); H.A. Jahn, ibid, *164,* 117 (1937).
207) W.D. Hobey and A.D. McLachlan, J.chem.Physics *33,* 1695 (1960); H.M. McConnell and A.D. McLachlan, ibid. *34,* 1 (1961).
208) H.M. McConnell, J.chem.Physics *34,* 13 (1961).
209) A. Carrington: Orbital Degeneracy and Spin Resonance in Free Radical-Ions. The Royal Institute of Chemistry, Lecture Series 1965.
210) E. Huckel, Z.Elektrochem. *43,* 752 (1937).
211) C.A. Coulson and H.C. Longuet-Higgins, Proc.Roy.Soc. *A 191,* 39 (1947); C.A. Coulson, Proc.physic.Soc. (London) *A 65,* 933 (1952); H.C. Longuet-Higgins and R.G. Sowden, J.chem.Soc. (London) *1952,* 1404.
212) a) E. Vogel and H.D. Roth, Angew.Chem. *76,* 145 (1964); E. Vogel and W.A. Böll, ibid, *76,* 784 (1964); b) E. Vogel, W. Meckel, and W. Grimme, ibid. *76,* 785 (1964); E. Vogel et al., unpublished work.
213) V. Boekelheide and J.B. Phillips, J.Amer.chem.Soc. *85,* 1545 (1963); Proc.Natl.Acad.Sci.U.S. *51,* 550 (1964).
214) F. Sondheimer and Y. Gaoni, J.Amer.chem.Soc. *82,* 5765 (1960).
215) N.M. Atherton, personal communication.
216) A.W. Hanson, Acta crystallogr. *18,* 599 (1965).
217) O. Bastiansen and O. Hassel, Acta chem.Scand. *3,* 209 (1949); I.L. Karle, J.chem.Physics *20,* 65 (1952); M. Marain and D. Saksena, Nature *165,* 723 (1950).
218) R.A. Raphael: Cyclooctatetraene, in D. Ginsburg: Non-Benzenoid Aromatic Compounds. Interscience, New York 1959.
219) T.J. Katz, J.Amer.chem.Soc. *82,* 3784, 3785 (1960).
220) J.P. Colpa, International Conference on Chemical Aspects of Electron Spin Resonance, Cardiff, Wales 1966.
221) R.G. Lawler, J.R. Bolton, G.K. Fraenkel, and T.H. Brown, J.Amer.chem. Soc. *86,* 520 (1964).
222) E.A. Halevi: Secondary Isotope Effects, in S.G. Cohen, A. Streitwieser, and

R.W. Taft: Progress in Physical Organic Chemistry. Interscience, New York 1963, Vol. I.

223) A. Carrington, H.C. Longuet-Higgins, R.E. Moss, and P.F. Todd, Mol.Physics *9.* 187 (1965).

224) M. Karplus, R.G. Lawler, and G.K. Fraenkel, J.Amer.chem.Soc. *87,* 5260 (1965).

225) K.W. Bowers and F.D. Greene, J.Amer.chem.Soc. *85,* 2331 (1963).

226) K.W. Bowers, G.J. Nolfi, and F.D. Greene, J.Amer.chem.Soc. *85,* 3707 (1963).

227) M.T. Jones, J.Amer.chem.Soc. *88,* 174 (1966).

228) F. Gerson, E. Heilbronner, and J. Heinzer, Tetrahedron Letters (London) *19,* 2095 (1966).

229) W.D. Phillips, J.C. Rowell, and S.I. Weissman, J.chem.Physics *33,* 626 (1960).

230) P.H. Rieger, I. Bernal, and G.K. Fraenkel, J.Amer.chem.Soc. *83,* 3918 (1961).

231) E. de Boer and J.P. Colpa, J.physic.Chem. *71,* 21 (1967).

232) D.H. Levy and R.L. Myers, J.chem.Physics *43,* 3063 (1965).

233) G.W. Griffin and L.I. Peterson, J.Amer.chem.Soc. *84,* 3398 (1962): P.A. Waitkus, E.B. Sanders, L.I. Peterson, and G.W. Griffin, ibid. *89,* 6318 (1967).

234) G. Köbrich and H. Heinemann, Angew.Chem. *77,* 590 (1965).

235) H. Hopff and A. Wick, Helv.chim.Acta *44,* 19, 380 (1961); H. Hopff and A. Gati, ibid. *48,* 1289 (1965).

236) F. Gerson, Helv.chim.Acta *47,* 1941 (1964).

237) F. Gerson, G. Köbrich, and E. Heilbronner, Helv.chim.Acta *48,* 1525 (1965).

238) E. Weltin, F. Gerson, J.N.Murrell, and E. Heilbronner, Helv.chim.Acta *44,* 1400 (1961).

239) L.H. Piette, P. Ludwig, and R.N. Adams, J.Amer.chem.Soc. *83,* 3909 (1961); *84,* 4212 (1962).

240) K.L. McEwen, J.chem.Physics *32,* 1801 (1960).

241) C. Lagercrantz and M. Yhland, Acta chem.Scand. *16,* 1807 (1962).

242) a) T.R. Tuttle and S.I. Weissman, J.chem.Physics *25,* 189 (1956); b) T.R. Tuttle, ibid. *32,* 1579 (1960).

243) K. Markau and W. Maier, Z.Naturforsch. *16a,* 636 (1961).

244) H.L. Strauss and G.K. Fraenkel, J.chem.Physics *35,* 1738 (1961).

245) M.R. Das and B. Ventakaraman, J.chem.Physics *35,* 2262 (1961); Bull.Coll.Amp.Eidhoven *1962,* 21.

246) T.N. Tozer and L.D. Tuck, J.chem.Physics *38,* 3035 (1963).

247) E.W. Stone and A.H. Maki, J.Amer.chem.Soc. *87,* 454 (1965).

248) N. Hirota, J.chem.Physics *37,* 1884 (1962).

249) E. de Boer, Recueil Trav.chim.Pays-Bas *84,* 609 (1965).

250) T.R. Tuttle, Ph.D. Thesis, Washington-University 1959; quoted in 91).

251) H. Nishigushi, Y. Nakai, K. Nakamura, K. Ishizu, Y. Deguchi, and H. Takaki, a) J.chem.Physics *40,* 241 (1964); b) Mol.Physics *9,* 153 (1965).

252) A.C. Aten, J. Dieleman, and G.J. Hoijtink, Disc. Faraday Soc. *29,* 182 (1960); J. Dieleman (1962) and K.H.J. Buschow (1963), Dissertations, Vrije Universiteit, Amsterdam; K.H.J. Buschow, J. Dieleman, and G.J. Hoijtink, J.chem.Physics *42,* 1993 (1965).
253) Landolt-Börnstein: Zahlenwerte und Funktionen. 6th edit., Springer, Heidelberg 1959, Vol. II, part 6; J.L. Down, J. Lewis, B. Moore, and G. Wilkinson, J.chem.Soc. (London) *1959,* 3767.
254) G. Wittig and R. Polster, Liebigs Ann.Chem. *599,* 1 (1956); G. Wittig and F. Stahnecker, ibid. *605,* 69 (1957).
255) N.M. Atherton and A.E. Goggins, a) Mol.Physics *8,* 99 (1964); b) Trans.Faraday Soc. *61,* 1399 (1965).
256) J. dos Santos-Veiga and A.F. Neiva Correia, Mol.Physics *9,* 395 (1965).
257) C.A. McDowell and K.F.G. Paulus, Canad.J.Chem. *43,* 224 (1965).
258) E.A.C. Lucken, J.chem.Soc. (London) *1964,* 4234.
259) G.R. Luckhurst and L.E. Orgel, Mol.Physics *7,* 297 (1963-64).
260) D.R. Eaton, Inorg.Chem. *3,* 1268 (1964).
261) E. König, Z.Naturforsch. *19a,* 1139 (1964).
262) A.H. Reddoch, a) J.chem.Physics *41,* 444 (1964); b) ibid. *43,* 225 (1965).
263) A. Carrington, Mol.Physics *5,* 425 (1962).
264) Ch.S. Johnson, Jr.: Chemical Rate Processes and Magnetic Resonance, in J.S. Waugh: Advances in Magnetic Resonance. Academic Press, New York 1965, Vol. I.
265) F. Gerson, E. Heilbronner, A.H.Reddoch, D.H. Paskovich, and N.C. Das, Helv.chim.Acta, *50,* 813 (1967).
266) J.R. Bolton, A. Carrington, and P.F. Todd, Mol.Physics *6,* 169 (1963).
267) S.I. Weissman, J.Amer.chem.Soc. *80,* 6462 (1958).
268) J.E. Harriman and A.H. Maki, J.chem.Physics *39,* 778 (1963).
269) A. Carrington and H.C. Longuet-Higgins, Mol.Physics *5,* 447 (1962).
270) J.H. Freed and G.K. Fraenkel, J.chem.Physics *39,* 326 (1963); *40,* 1815 (1964).
271) V. Boekelheide and G.K. Vick, J.Amer.chem.Soc. *78,* 653 (1956).
272) R.L. Letsinger and J.A. Gilpin. J.org.Chemistry *29,* 243 (1964).
273) V. Boekelheide, W.E. Langeland, and Chu-Tsin Liu, J.Amer.chem.Soc. *73,* 2432 (1951).
274) R.J. Windgassen Jr., W.H. Saunders Jr., and V. Boekelheide, J.Amer.chem.Soc. *81,* 1459 (1959).
275) E. de Boer and C. MacLean, Mol.Physics *9,* 191 (1965).
276) K.H. Hausser, H. Brunner, and J.C. Jochims, Mol.Physics *10,* 253 (1966).
277) J. dos Santos-Veiga, W.L. Reynolds, and J.R. Bolton, J.chem.Physics *44,* 2214 (1966).

278) W.H. Bruning, G. Henrici-Olivé, and S. Olivé. Z.physik.Chem. (Frankfurt/M) *47,* 114 (1965).
279) A.H. Reddoch, personal communication.
280) T.A. Claxton, W.M. Fox, and M.C.R. Symons, Trans Faraday Soc. *63,* 2570 (1967).
281) D. Kusch and H. Taub, Phys.Reviews *75,* 1477 (1949).
282) R.S. Mulliken, J.Amer.chem.Soc. *72,* 600 (1950); *74,* 811 (1952).
283) W.M. Gulick and D.H. Geske, J.Amer.chem.Soc. *87,* 4049 (1965); *88,* 4119 (1966); B.L. Silver, Z. Luz, and C. Eden, J.chem.Physics *44,* 4258 (1966); M. Brose, Z. Luz, and B.L. Silver, ibid. *46,* 4891 (1968).
284) D.H. Levy and R.L. Myers. J.chem.Physics *44,* 4177 (1966).
285) T.M. McKinney and D.H. Geske, J.Amer.chem.Soc. *87,* 3013 (1965).
286) G.A. Russel, E.T. Strom, E.R. Talaty, and S.A. Weiner, J.Amer.chem.Soc. *88,* 1998 (1966).
287) G.A. Russell, personal communication.
288) G.A. Russell and R.D. Stephens, J.physic.Chem. *70,* 1320 (1966).
289) G.A. Russell and E.T. Strom, J.Amer.chem.Soc. *86,* 744 (1964).
290) G.A. Russell, R.D. Stephens, and E.R. Talaty, Tetrahedron Letters *1965,* 1139; E.T. Strom, G.A. Russell, and R.D. Stephens, J.physic.Chem *69,* 2131 (1965).
291) G.A. Russell and E.R. Talaty, J.Amer.chem.Soc. *86,* 5345 (1964); *87,* 4867 (1965); Science *148,* 1217 (1964).
292) E.T. Strom and G.A. Russell, J.chem.Physics *41,* 1514 (1964).
293) J.E. Bloor, B.R. Gilson, and P.N. Daykin, J.physic.Chem. *70,* 1457 (1966).
294) C. Lagercrantz, K. Torssell, and S. Wold, Arkiv för Kemi *25,* 567 (1966).
295) A. Carrington and G.R. Luckhurst, Mol.Physics *8,* 401 (1964).
296) H.R. Falle and G.R. Luckhurst, Mol. Physics *11,* 299 (1966); *12,* 493 (1967); S.H. Glarum and J.H. Marshall, J.chem.Physics *44,* 2884 (1966).
297) R.W. Brandon and E.A.C. Lucken, J.chem.Soc. (London) *1961,* 4273; A. Fairbourn and E.A.C. Lucken, ibid. *1963,* 258.
298) J.E. Wertz and J.L. Vivo, J.chem.Physics *23,* 2441 (1955).
299) M. Plato, Z.Naturforsch. *22a,* 119 (1967).
300) F.C. Adam and S.I. Weissman, J.Amer.chem.Soc. *80,* 1518 (1958).
301) M. Bersohn and J.C. Baird: An Introduction to Electron Paramagnetic Resonance. Benjamin, New York 1966.
302) A. Carrington and A.D.McLachlan: Introduction to Magnetic Resonance. Harper and Row, New York 1967.
303) B.H.J. Bielski and J.M. Gebicki: Atlas of Electron Spin Resonance Spectra. Academic Press, New York and London 1967.

304) P.B. Ayscough: Electron Spin Resonance in Chemistry. Methuen, London 1967.
305) V. Boekelheide, W.E. Langeland, and Chu. Tsin-Liu, J. Amer. chem. Soc. *73,* 2432 (1951).
306) K. Hafner and G. Schneider, Liebigs Ann.Chem. *672,* 194 (1964); K. Hafner et. al., unpublished work.
307) A.F. Neiva-Correia, Thesis, Universiteit van Amsterdam 1967.
308) G.K. Fraenkel, J. physic.Chem. *71,* 139 (1967).
309) D.H. Geske and G.R. Padmanabhan, J. Amer. chem. Soc. *87,* 1651 (1965).
310) J.R. Morton, Chem. Reviews *64,* 453 (1964).
311) H.M. McConnell, C. Heller, T. Cole, and R.W. Fessenden, J. Amer.chem.Soc. *82,* 766 (1960).
312) G.W. Canters and E. de Boer, Mol. Physics *13,* 395 (1967).
313) K. Nakamura, Bull.chem.Soc.Japan *40,* 1019 (1967).
314) A.M. Hermann, A. Rembaum, and W.R. Carper, J.physic.Chem. *71,* 2661 (1967).
315) C.-Y. Ling and J. Gendell, J.chem.Physics *46,* 400 (1967); *47,* 3475 (1967); R.J. Faber and G.K. Fraenkel, ibid. *47,* 2462 (1967).
316) H.J. Shine and P.D. Sullivan, J. physic. Chem. *72,* 1390 (1968); P.D. Sullivan, J. Amer. chem. Soc. *90,* 3618 (1968).
317) M.C.R. Symons, J. physic. Chem. *71,* 172 (1967).
318) N. Hirota, J. physic. Chem. *71,* 127 (1967).
319) N. Hirota and R. Kreilick, J. Amer. chem. Soc. *88,* 614 (1966); A.H. Crowley, N. Hirota and R. Kreilick, J. chem. Physics *46,* 4815 (1967).
320) J. Sinclair and D. Kivelson, J. Amer. chem. Soc. *90,* 5074 (1968).
321) G.K. Fraenkel, A. Kaplan, and J.R. Bolton, J. chem. Physics *42*, 955 (1965).
322) R.W. Fessenden and R.H. Schuler, J. chem. Physics *43,* 2704 (1965).
323) E.T. Kaiser and L. Kevan: Radical Ions. Wiley, New York 1968.
324) G.A. Russell: Semidione Radical Anions, In Ref. 323).
325) H.R. Blattmann, W.A. Böll, E. Heilbronner, G. Hohlneicher, E. Vogel, and J.-P. Weber, Helv. chim. Acta *49,* 2017 (1966).
326) M. Dobler and J.D. Dunitz, Helv. chim. Acta *48,* 1430 (1965).
327) P.R. Hindle, J. dos Santos-Veiga, and J.R. Bolton, J. chem. Physics *48,* 4703 (1968).
328) R.G. Lawler, J.R. Bolton, M Karplus, and G.K. Fraenkel, J. chem. Physics *47,* 2149 (1967).
329) E.E. van Tamelen and T.L. Burkoth, J. Amer. chem. Soc. *89,* 151 (1967).
330) R.S. Alger: Electron Paramagnetic Resonance. Interscience Publishers, New York 1968.
331) R.G. Lawler and G.K. Fraenkel, J. chem. Physics *49,* 1126 (1968).

Subject Index

E

F

G

H

I

K

L

M

N

O

P

Q